Exploration & Discovery

Treasures of the Yale Peabody Museum of Natural History

Yale Book Award

PRESENTED BY

SOUTH HADLEY HIGH SCHOOL

TO

LUKE PETROSKY

For Outstanding Personal Character and Intellectual Promise

MAY 31, 2018

Date

ASSOCIATION OF YALE ALUMNI

Exploration & Discovery

Treasures of the Yale Peabody Museum of Natural History

David K. Skelly
Thomas J. Near

Photography by Robert Lorenz

With a foreword by James Prosek

Peabody Museum of Natural History
Yale University
New Haven, Connecticut

Distributed by Yale University Press
NEW HAVEN AND LONDON

Yale

Support for this book has been provided by the O.C. Marsh Fellows Program at the Yale Peabody Museum of Natural History

Editors
Richard A. Kissel, *Director of Public Programs*
Rosemary Volpe, *Publications Manager*

Design
Sally H. Pallatto

Photography by Robert Lorenz
(unless otherwise noted)

Index
Judy Hunt

Printed by GHP Media, Inc.

ISBN 978-1-933789-05-7
Printed in the USA

Published by Peabody Museum of Natural History, Yale University
P. O. Box 208118, New Haven CT 06520-8118 USA
peabody.yale.edu

Distributed by Yale University Press
New Haven and London | yalebooks.com

Library of Congress Control Number: 2016935431

This paper meets the requirements of ANSI / NISO Z39.48-1992 (Permanence of Paper).

10 9 8 7 6 5 4 3 2 1

Contents

Tympanuchus cupido cupido
COLUMBIDAE
Ectopistes migratorius
Ectopistes migratorius
Passenger Pigeon 1900 / 1914
Ectopistes migratorius
N. America
PEABODY MUSEUM OF YALE UNIVERSITY
6982

Foreword

I can vividly remember what it felt like as a child to walk into the neo-gothic space and stare in awe at the model of a giant squid suspended from the ceiling. The didactic said these beasts fought with whales in the deep sea. Turning a corner around the front desk I was in the glorious dinosaur hall, craning my neck at a large fossil skeleton of a prehistoric reptile. It overwhelmed the spirit. To think a creature as big as a whale once walked on land?

One of the most famous natural history murals in the world, perhaps the standard by which all are measured, commanded the right side of the room—Rudolph Zallinger's masterpiece on the evolution of dinosaurs. In its beauty and richness of atmosphere, I feel this work obliterates all renderings in still or motion picture of that colorful world millions of years ago that we will never see.

Around another corner was the prehistoric horse, an Incan feather tapestry, a display of stone projectile points and ancient fishing traps. What did all these objects mean? At the time I didn't know exactly, and I still puzzle at how and why they fall together in the same house. I took in the diversity of natural forms and the human efforts to make sense of them. One might think that spending time here would encourage a child to become a biologist or anthropologist. On reflection I see that visits to the Peabody encouraged me to be an artist.

I still return to walk through the exhibits and to marvel at what evolution has fashioned—in the form of animals and plants, but also the human mind and what, exposed to vastly diverse experiences, it is capable of creating. Evolution made us, and we make paintings, decorated vessels, bejeweled adornments, samurai swords. I suppose it makes sense that all these things should be together. It is a museum of natural objects but even more so it is a museum of man and nature.

The Peabody is complemented by two major art museums on the other side of campus. But the natural history museum to me is special, in part because it is less specialized. It still lives in an earlier era where science and art were not classified as separate named entities, where objects found in nature, human ideas, and human creations lived side by side. There is something extraordinary in this. Perhaps this kind of amalgamated display, as was more common in the past, is also a glimpse of the future? Why can't a bird's nest and a reed basket live side by side? They are both the creations of deft weavers, who make things for function but also perhaps just for the sake of beauty. Such juxtapositions question the very foundation of art and of science as discreet disciplines—and whether humans are separate from nature or a part of it.

This tantalizing book offers a taste of the objects, curators, artists and scientists that have made the Peabody one of the great institutions of the world.

— James Prosek

The passenger pigeon—once the most abundant bird in North America, perhaps in the world—became extinct when the last remaining bird—"Martha"—died in captivity on September 1, 1914. The Peabody collection makes this and other lost species available for research today.

Ectopistes migratorius **specimens**
Ornithology collections
Division of Vertebrate Zoology
Class of 1954 Environmental Science Center
Yale University

Anguilliformes
NEMICHTHYIDAE
Anguilliformes
OPHICHTHIDAE

Preface

Yale's Peabody Museum of Natural History was founded in 1866, making it one of the oldest university-based museums. It is also one of the largest. When queried, visitors to the Museum's galleries will estimate that about 25% of the collections are on display. The real number is 0.04%. The Peabody's 13 million specimens arrayed across ten divisions represent, collectively, an enormous endeavor. It is more than reasonable to ask why. Why so many specimens and objects? What purpose do they serve? This book, and the exhibition on which it is based, offer one answer.

The fact is that material objects represent the most powerful foundation we possess to understand our world. The Peabody was created because Yale's first scientists recognized that they needed a museum to make pathbreaking discoveries. Their foresight paid off. The Museum has been the platform from which some of the best scientific minds have changed how we think about the planet and its inhabitants. That work continues today.

This book is organized into seven sections representing a sampling of the different ways that the Peabody Museum and its collections have been at the center of important discoveries and major scientific debates. While even the name "museum" can feel old, and museums can seem like they are both about the past and of the past, the stories we present in these pages emphasize over and over that the Peabody has always been about discovery and innovation. Just as they always have, the scientists and staff of the Peabody are aimed at rewriting our knowledge of the natural world and sharing it with audiences both on campus and around the world.

—David Skelly and Tom Near
February 2016

In addition to skeletal material, the Peabody's Division of Vertebrate Zoology has over 73,000 specimens of fishes, reptiles, amphibians, birds, and mammals stored in fluid, which also includes some that are cleared and stained, a process whereby a specimen is made transparent and the bones and cartilage are stained different colors. "Wet" specimens are useful to several different types of research, such as studies of feeding ecology, reproductive biology, developmental biology and parasitology, conservation genetics, and systematics.

Fluid collections
Division of Vertebrate Zoology
Class of 1954 Environmental Science Center
Yale University

1

Exploration, Discovery, and Understanding

From the bones of massive dinosaurs to cultural artifacts from across the globe, the collections of natural history are more than curiosities—they are the foundation for our understanding of the natural world.

In this book we invite you to explore the history and impact of the Yale Peabody Museum of Natural History. It is a story of exploration and discovery, and of finding our place in the natural world. We begin our story with an overview of the Museum and its history.

Opposite

1 This is the first microscope acquired by Yale College, purchased in 1735. Although Yale's first professor of science would not be hired for another seven decades, this singular instrument represents the foundation of science at Yale.

Culpeper/Loft Double Microscope

2 Artist's sketch of New Haven fortifications constructed in 1676 in response to King Philip's War.

"New Haven Palisaded or Fortified" From *A Pictorial History of Raynham and Its Vicinity* by Charles Hervey Townshend, 1900

Birth of a University

Yale's beginnings are rooted in the 1701 passage of An Act for Liberty to Erect a Collegiate School, which created a small school in what is now Clinton, Connecticut. Fifteen years later the school moved to New Haven.

In 1718 Cotton Mather contacted Elihu Yale, an American-born merchant who had acquired his fortune with the British East India Company. His purpose? Obtaining funds for the construction of a new building at the Collegiate School.

3 **Portrait of Gov. Elihu Yale (1649–1721)**
Oil on canvas (1717)
Artist, Enoch Seeman the younger, British (1694–1744)

Yale offered 417 books, a portrait of King George I, and nine bales of goods that were sold for the handsome sum of 800 British pounds.

The school was named Yale College in his honor.

4 "A View of the Buildings of Yale College at New Haven" (1807)
Lithograph
Artist, Amos B. Doolittle (1754–1832)

The Brick Row campus plan—developed in 1792 by John Trumbull (a one-time aide to General George Washington) and James Hillhouse—made Yale the first planned college campus in America. The center building stands today as Connecticut Hall, which possibly housed the microscope shown opposite page 1.

Yale's First Scientist

The scientific revolution that blossomed during the Age of Enlightenment came late to America. Yale College had very little science in its curriculum during its first century.

But in 1795 Yale appointed as its president Timothy Dwight, an energetic proponent of scientific inquiry and founder of the Connecticut Academy of Arts and Sciences.

In 1801—the year that Thomas Jefferson took office as the third U.S. President—Dwight offered family friend and recent Yale graduate Benjamin Silliman (1779–1864) the first professorship of science at Yale.

Silliman had a profound influence on the birth of science in America and at Yale. As professor of "chymistry" and natural history, Silliman was Yale's first scientist. He was expected to teach not only chemistry but also botany, geology, mineralogy, and zoology.

With great energy and ambition, Silliman prepared for his new position by studying abroad for two years in England and Scotland. On his return in 1806, Silliman was well-versed in many scientific disciplines, bringing science to American shores.

Silliman's assortment of minerals and other natural objects would ultimately form the foundation of the Peabody Museum's collection. His work at Yale would also attract the interest of students who would contribute directly to the founding of the Museum.

B Silliman

***5* Benjamin Silliman (1779–1864)**
Oil on ivory, miniature (ca. 1815)
Artist, Nathaniel Rogers (1788–1844)

***6* Mineral (Hematite) purchased by Benjamin Silliman in London (1805)**

Among Benjamin Silliman's long list of contributions to science is the *American Journal of Science.* Created by Silliman in 1818, it is America's longest-running scientific journal, still published today.

7 **Kelp ashes in glass jar belonging to Benjamin Silliman ca. 1817**

8 **The inaugural issue of the *American Journal of Science,* 1818**

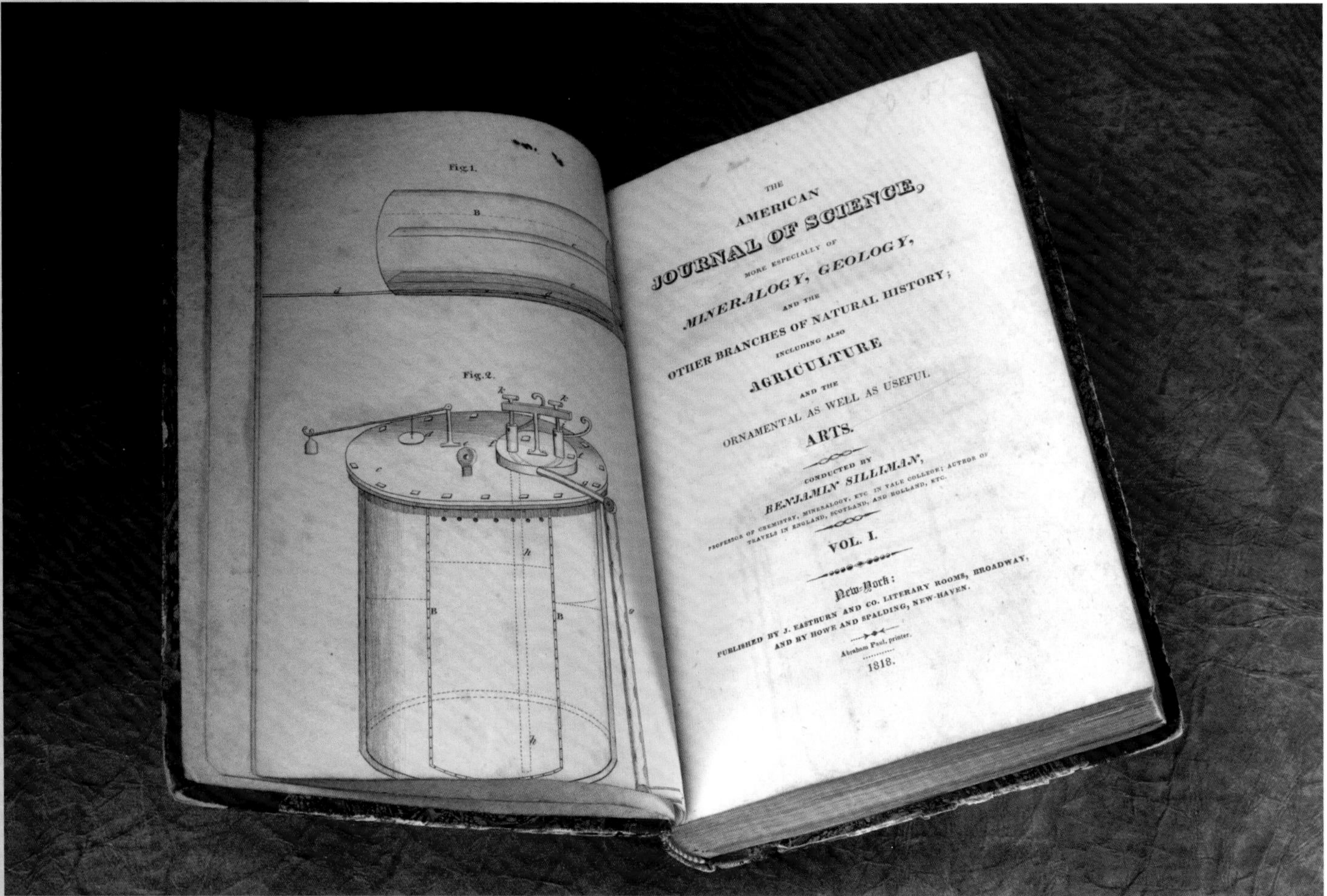

At 6:30 on the morning of December 14, 1807, a blazing fireball about two-thirds the size of the moon was seen traveling southward above New England. Three loud explosions were heard over the town of Weston, Connecticut.

Shortly after, Silliman and his Yale colleague James Kingsley traveled to Weston and secured "a considerable number of specimens."

On December 29 they published an account of the event—the first recorded meteorite fall in the New World. In March 1809 Silliman published the first chemical analysis of the meteorite—the first in the country and among the first in the world. This work brought prominence to American science and established Yale's meteorite collection—the oldest in the United States.

9 The Weston meteorite from the December 14, 1807 recorded fall. Silliman's published account and chemical analysis pointed to an extraterrestrial origin.

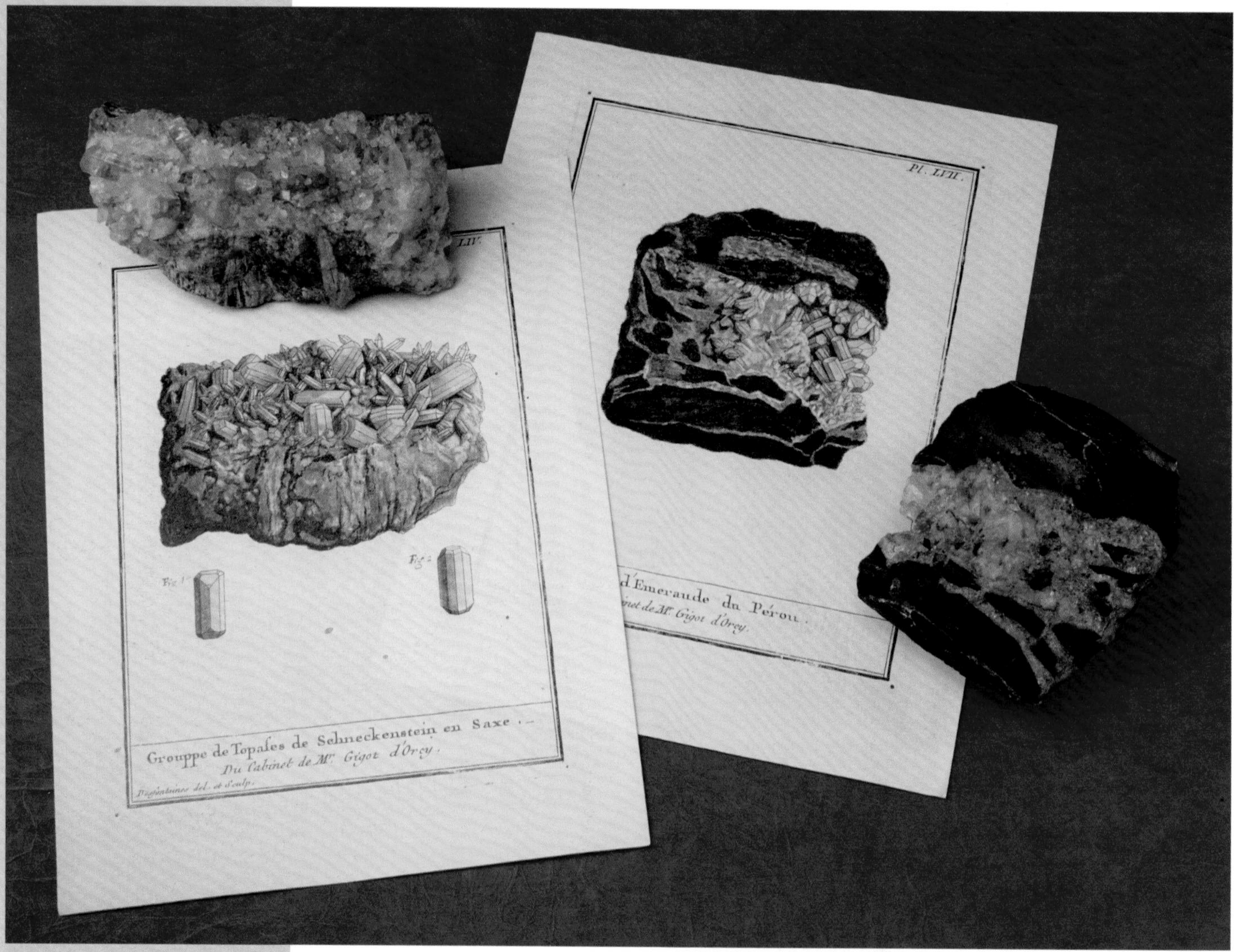

10* Beryl, variety emerald (right) and topaz from the Gigot d'Orcy Collection, with facsimiles of illustrated plates for the same specimens published in the 18th century series *Histoire Naturelle.

While acquiring the knowledge he needed for teaching, Silliman was also acquiring materials. Only 20 years later Yale's mineral collection would be the largest and most important in the United States.

In addition to the specimens he collected personally, Silliman oversaw several important acquisitions, including the Perkins Cabinet in 1807 and the Gibbs Cabinet in 1825.

Elements of the Gibbs Cabinet were initially assembled in 18th century pre-Revolutionary France by Jean Baptiste François Gigot d'Orcy, by Grigori Razumovsky in Switzerland, and by Jaques-Louis, comte de Bournon in Britain. Shown here are original specimens from this collection.

Science Is Established at Yale

Reacting to the expedition of the HMS *Beagle*, on which a young naturalist named Charles Darwin had sailed, the U.S. Congress funded the United States Exploring Expedition.

Focused on the Pacific, the four-year expedition launched in 1838. Invited as a scientific participant was Yale graduate and student of Benjamin Silliman—James Dwight Dana (1813–1895).

As the Silliman Professor of Natural History, Dana is among Yale's most notable scientific figures. His contributions to zoology and geology—in particular the classification of minerals he outlined in *A System of Mineralogy*—are still in use today.

On the heels of the establishment of the Yale (later Sheffield) Scientific School in 1847, Dana was offered a professorship and quickly became an eloquent voice for the expansion of science at Yale. He called for more laboratories and lecture halls, and for a museum in which "no one could walk through its halls without profit."

In 1866, the Yale Peabody Museum of Natural History would be founded.

11 **James Dwight Dana (1813–1895)**
Oil on canvas (1858)
Artist, Daniel Huntington (1816–1906)

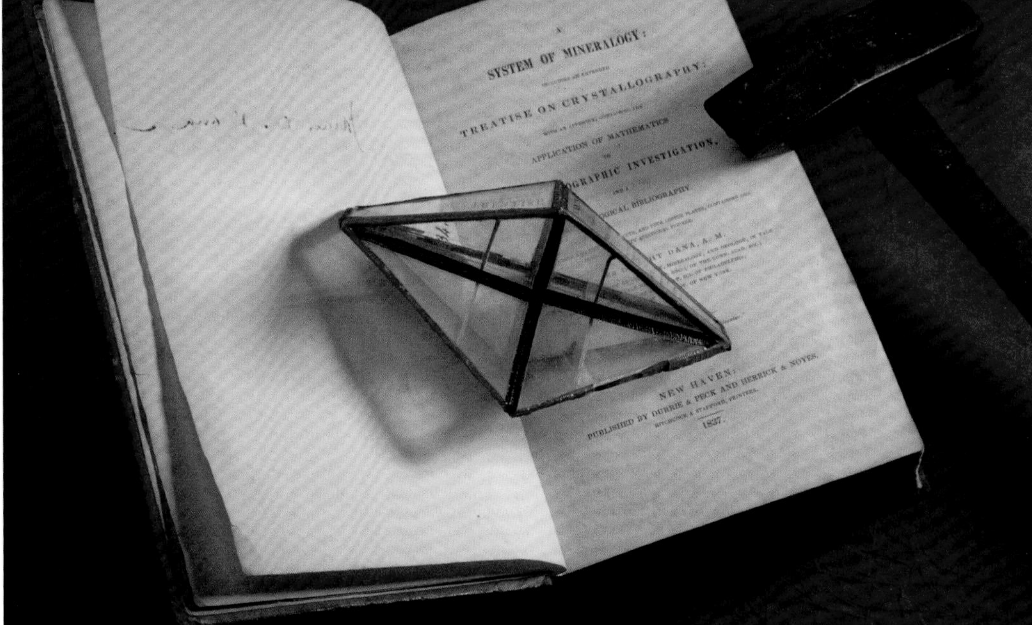

12 **J.D. Dana's rock hammer, a glass crystal model (calcite twin) that he made for teaching students, and his personal signed copy of *A System of Mineralogy*.**

***A System of Mineralogy*, first edition (1837)**
Author, James Dwight Dana

Peabody's New Museum

In August 1856—the same year that saw professor James Dwight Dana call for the creation of a museum of natural history at Yale—aspiring freshman Othniel Charles Marsh (1831–1899) took (and aced) Yale's admission exam.

Inspired by the work of Silliman, Marsh was deeply interested in science and its interpretation of the world.

Soon after his graduation from Yale's Sheffield Scientific School in 1862, Marsh begins discussions with his uncle—international financier George Peabody—for a donation to design and build a natural history museum at Yale, under the guidance of Silliman and Dana.

A deed of gift for $150,000 was handed to Marsh in 1866, marking the founding of Yale's Peabody Museum of Natural History.

The year 1866 saw not only the founding of the Peabody Museum but also the appointment of Marsh as Yale's professor of paleontology, the first such appointment in the United States.

13 **Othniel Charles Marsh (1831–1899), ca. 1856**

In 1867 Marsh was also appointed one of the Museum's first curators—along with geologist George J. Brush and zoologist Addison E. Verrill—and assumed the (unofficial) directorship of the Peabody.

14 **George Peabody (1795–1869)**
Oil on canvas (date unknown)
Artist unknown

ORIGINAL DOCUMENTS

OF

THE PEABODY MUSEUM OF NATURAL HISTORY,

YALE UNIVERSITY.

MR. PEABODY'S LETTER.

NEW YORK, Oct. 22, 1866.

To Professor JAMES D. DANA, Hon. JAMES DIXON, Hon. ROBERT C. WINTHROP, Professor BENJAMIN SILLIMAN, Professor GEORGE J. BRUSH, Professor OTHNIEL C. MARSH, and GEORGE PEABODY WETMORE, Esq.

GENTLEMEN :—With this letter I enclose an instrument giving to you one hundred and fifty thousand dollars ($150,000), in trust for the foundation and maintenance of a MUSEUM OF NATURAL HISTORY, especially of the departments of Zoology, Geology, and Mineralogy, in connection with Yale College.

I some years ago expressed my intention of making a donation to this distinguished institution, and convinced as I am of the importance of the natural sciences, and of the increasing interest taken in their study, it now affords me great pleasure to aid in advancing these departments of knowledge.

The rapid advance which natural science is now making renders it necessary to provide for the future requirements of such a museum, as well as its present wants, and I trust that the portion of the fund designed for this purpose will be found sufficient.

On learning of your acceptance of this trust, and of the assent of the President and Fellows of Yale College to its conditions, I shall be prepared to pay over to you the sum I have named, and I may then have some additional suggestions to make, in regard to the general management of the trust.

Confident that under your direction this trust will be faithfully and successfully administered,

I am, with great respect, your obedient servant,

GEORGE PEABODY.

15 **The Deed of Gift that founded the Yale Peabody Museum of Natural History in 1866.**

16 **O. C. Marsh (standing, center) and the expedition crew of the Yale College Scientific Expedition of 1870 at their field camp near Fort Bridger, Wyoming.**

Marsh's early research involved four Yale College Scientific Expeditions. From 1870 to 1873, Marsh and his students explored the American West in search of fossils, primarily those of ancient mammals.

Meanwhile, Verrill's research covered a breadth of topics, from botany to parasitology—as evidenced by these tapeworms, removed from Yale College students on March 18, 1896.

17 **Addison Emery Verrill (1839–1926)**
Oil on canvas
1910
Artist, John Henry Niemeyer (1839–1932)

18 **Beef tapeworm (large jar) from a Yale College student**
Taenia saginata
Acquired by A. E. Verrill, 1896

19 **George Jarvis Brush (1831–1912)**
Oil on canvas
(date unknown)
Artist, Harry Ives Thompson (1840–1906)

20 **Architectural rendering of the first Peabody Museum building, as originally designed Watercolor and graphite on paper (ca. 1873) Artist, J. Cleaveland Cady (1837–1919)**

The site on the southwest corner of Elm and High streets was chosen for the Peabody Museum. Construction began in the summer of 1874 and was completed in 1876. The Museum opened to the public the same year.

Although the architect's original vision called for a large and imposing edifice, only one wing of the design was ever built.

21 **The first Peabody Museum building had only enough room to display the pelvis and hind legs of *Brontosaurus,* mounted by long-time preparator Hugh Gibb (posing next to the left leg, ca. 1902).**

22 **Undated photograph of the first Peabody Museum building on Elm and High streets in New Haven.**

23 **Exhibit display of the Otisville mastodon and other skeletons in the first Peabody Museum.**

24 Peabody paleobotanist George R. Wieland (1865-1953) stands beside the mounted skeleton of *Archelon ischyros* (completed in 1907 and clearly missing a rear flipper, lost to a predator in life) in the first Peabody Museum building, on November 3, 1914.

Collected from the American West, tons of fossils were transported by rail to New Haven. Here, Marsh described and named such iconic dinosaurs as *Allosaurus, Stegosaurus, Triceratops, Apatosaurus*, and the "thunder lizard" *Brontosaurus*.

But Marsh was not the only scientist interested in America's extinct fauna. Edward Drinker Cope of Philadelphia was just as ambitious. Cope and Marsh's early friendship soon soured, igniting the "Bone Wars" of the late 19th century.

_25_ O.C. Marsh's international renown is seen in this cartoon published in _Punch_.

"Professor Marsh's Primeval Troupe"
***Punch* Magazine (1890)**
Artist, possibly
Edward Tennyson Reed (1860–1933)

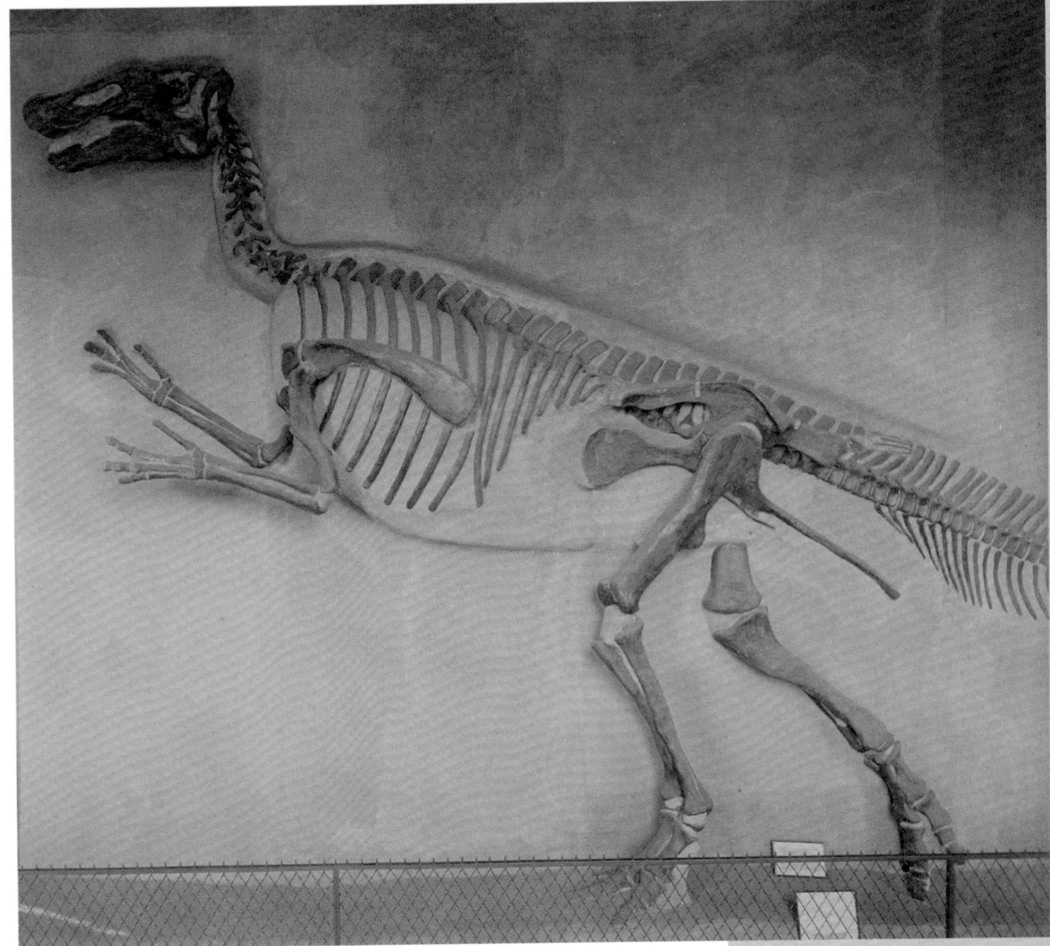

_26_ In 1901, this mount of _Claosaurus_ (now _Edmontosaurus_), the first large panel mount ever attempted for a dinosaur, was a sensation and made the front page of _The New York Times_.

Opposite

27 **This original 1867 Peabody Museum archaeology and ethnology catalogue lists this Egyptian vase as object 1 in the Museum's anthropology collection.**

Vase
Ptolemaic period (300–100 B.C.)
Island Elephantine, near the second cataract of the Nile

28 **Puma effigy stone mortar**
Late Horizon, Inca
Cuzco, Peru

Below, inset

29 **Hiram Bingham at the main camp in Peru, in September 1912.**

Not long after Marsh's death in 1899, Anthropology was added as a new division of the Museum's scientific research in 1902.

In 1911, Hiram Bingham III (1875–1956) led the first of three expeditions to Peru, bringing the Inca site of Machu Picchu to the world's attention.

Hiram Bingham

30 **The Inca ruins at Machu Picchu, from a series of hand-colored slides from the Yale Peruvian Expedition by photographer Harry Ward Foote, a Yale chemistry professor who served as the expedition collector and naturalist.**

The House that *Brontosaurus* Built

The first Peabody Museum building opened to the public in 1876, but its capacity was soon strained by the huge dinosaur bones that Marsh's collectors were sending back from the American West.

In 1917 it was demolished to make way for a new Yale dormitory complex, the Harkness Quadrangle and what became Saybrook College.

As plans for a new Museum building were being developed, construction was delayed by World War I. The collections were in nearly inaccessible storage for seven years.

Supported by a donation of $650,000 from Margaret Olivia Slocum Sage, property for the new Peabody Museum building was purchased at the corner of Whitney Avenue and Sachem Street, displacing a set of tennis courts then sited there.

The cornerstone was laid during commencement week, on June 18, 1923.

Dedicated on December 29, 1925, the new Peabody Museum opened to the public early the following year. Director Richard Swann Lull's (1867–1957) vision—to display the story of evolution "from the amoeba to man," as stated in *The New York Times*—was considered an overwhelming success.

31 **Architectural drawing of the current Peabody Museum building, showing a proposed extension that was never built (1930) Artist, Charles Z. Klauder (1872–1938)**

32 **The current Peabody Museum under construction in 1924.**

33 **Postcard showing the Yale Tennis Court at Whitney Avenue and Sachem Street in New Haven. (date unknown)**

34 **Dedication of the current Peabody Museum building in 1925, attended by more than 800 people, including members from eight scientific societies.**

35 **Richard Swann Lull working on figures for the Jurassic diorama.**

July 10, 1925

Clarence Darrow, Esq.,
Dayton, Tenn.

Dear Sir:

Your telegram of this morning to Professor Lull was received here and retelegraphed to him at Cloyne Court, Berkeley, California, where he is delivering two lecture courses in the summer school of the University. You may therefore expect a reply from him within a reasonable time, but I doubt very much his being able to attend the trial.

In my opinion Professor Lull is entirely in sympathy with the defense and would stand back of any statements made in Organic Evolution (Macmillan, 1917) and The Ways of Life (Harper, 1925).

Very truly yours,

MAT:G

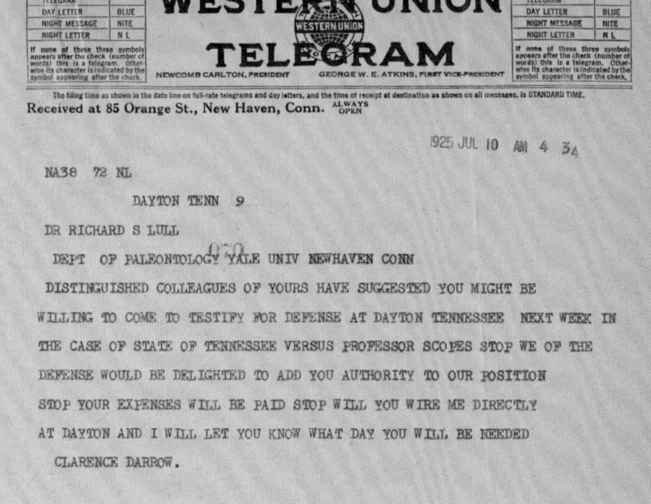

WESTERN UNION TELEGRAM

NEWCOMB CARLTON, PRESIDENT GEORGE W. E. ATKINS, FIRST VICE-PRESIDENT

Received at 85 Orange St., New Haven, Conn. ALWAYS OPEN

1925 JUL 10 AM 4 34

NA38 72 NL

DAYTON TENN 9

DR RICHARD S LULL

DEPT OF PALEONTOLOGY YALE UNIV NEWHAVEN CONN

DISTINGUISHED COLLEAGUES OF YOURS HAVE SUGGESTED YOU MIGHT BE WILLING TO COME TO TESTIFY FOR DEFENSE AT DAYTON TENNESSEE NEXT WEEK IN THE CASE OF STATE OF TENNESSEE VERSUS PROFESSOR SCOPES STOP WE OF THE DEFENSE WOULD BE DELIGHTED TO ADD YOU AUTHORITY TO OUR POSITION STOP YOUR EXPENSES WILL BE PAID STOP WILL YOU WIRE ME DIRECTLY AT DAYTON AND I WILL LET YOU KNOW WHAT DAY YOU WILL BE NEEDED

CLARENCE DARROW.

As the Scopes Trial raged in Tennessee—fueling a national debate on the teaching of evolution within public schools—Lull led the design for the new Museum's displays.

Lull's reputation as an educator and supporter of evolution was cemented with his appearance on the cover of the June 1, 1925 issue of *Time* magazine. Within its pages, the magazine praised Lull for his efforts.

Yale president James Rowland Angell encouraged Peabody officials to produce exhibits for use by local public schools. In line with this charge, Lull led the creation of displays that would illustrate evolutionary history as a spatially arranged walk through time—the first exhibition plan of its kind in any museum.

36, 37 **The July 10, 1925 telegram from defense attorney Clarence Darrow asking Lull to testify for the defense at the Tennessee trial of teacher John Scopes, and the Peabody's letter responding that Lull was lecturing in California that summer.**

In 1931 the Museum's Great Hall would finally serve its intended purpose with the installation of Marsh's gigantic *Brontosaurus,* adding to the Museum's story of evolution through the ages.

38 **Mounting of Marsh's *Brontosaurus* skeleton in 1930.**

39 **R. S. Lull and preparator Hugh Gibb stand beneath the completed *Brontosaurus.***

With the new Peabody building dedicated and its new displays on view, the Museum's collections continued to grow. In 1930, Harry Payne Bingham donated an entire collection from three oceanographic expeditions—conducted from 1925 to 1927—to Yale.

Specimens from the Bingham Oceanographic Expeditions included fishes and invertebrates from the Atlantic and Pacific oceans, ultimately adding a valuable collection of modern species to the Museum's holdings.

40 **This painting of barberfishes, surgeonfish, rock croaker, and spotted jawfish is part of a series done for the Bingham Oceanographic Expeditions.**

Untitled (1927)
Oil on canvas
Artist, Wilfrid Swancourt Bronson (1894–1985)

41 **This depiction of a dragonfly painted by Robert Bruce Horsfall (1869–1948) for the Peabody's former Carbonaceous Forest diorama is said to have inspired the career of Paul MacCready, the creator of the human-powered aircraft *Gossamer Condor.***

In the museum world, the Yale Peabody Museum's dioramas are considered masterpieces. A diorama combines three-dimensional foreground material with a curved background wall and domed ceiling to tell the story of an ecosystem. We accept the diorama as casually as if it were a window into the natural world.

Painters J. Perry Wilson (1889–1976) and Francis Lee Jaques (1887–1969), and the Peabody's chief preparator Ralph C. Morrill (1902–1996) who created the foregrounds, were the artists primarily responsible for the exquisite Peabody Museum dioramas. For several years beginning in October 1944, Perry Wilson, on leave from the American Museum of Natural History, created several of the Museum's dioramas, the largest being the 35-foot-long Coastal Region.

Wilson, trained in architecture, used his own unique technique to mathematically transfer an image onto a surface of any shape to create the illusion of space and distance. The Wilson method enabled him to avoid the inaccuracies inherent in freehand drawing. Generations of visitors have spent countless hours trying in vain to determine where foreground becomes background in a Wilson–Morrill diorama.

42, 43 **J. Perry Wilson's model (inset) of the Timber Line diorama (above) of bighorn sheep in the Canadian Rockies.**

44 **Early sketch, *The Age of Mammals* (1953)**
Graphite on paper
Artist, Rudolph F. Zallinger (1919–1995)

In 1941 Rudolph F. Zallinger (1919–1995), a senior at Yale's School of Fine Arts, agreed to illustrate seaweed for Museum director Albert E. Parr. One year later, Zallinger was tasked with illustrating prehistoric life in the Museum's Great Hall.

After a six-month crash course in animal and plant life—and the year-long production of an egg tempera study—Zallinger began work on the 110-foot-long piece. Using a dry fresco technique, four years of work would result in a brilliant depiction of 300 million years of evolution—*The Age of Reptiles.*

Zallinger received the Pulitzer traveling art scholarship in 1949 for his mural, and the studies for his *The Age of Reptiles* and *The Age of Mammals* murals appeared on the cover of *Life* magazine in 1953.

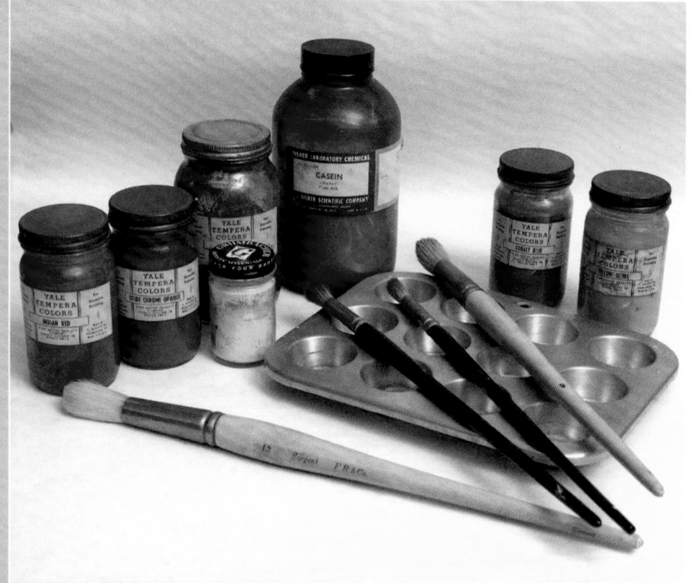

45 **Pigments, sable brushes, casein, and "palette" used by R. F. Zallinger to paint both *The Age of Reptiles* and *The Age of Mammals* murals.**

46 **Early sketch, *The Age of Reptiles* (1942)**
Graphite on paper
Artist, Rudolph F. Zallinger (1919–1995)

A World Leader

47 **The area of the quarry site in southern Montana, where in late August 1964 the fossils of *Deinonychus* were first discovered.**

With the second Peabody Museum building established as a Yale and New Haven landmark, its staff and researchers continued to expand the exhibition halls and build the scientific collection.

While combing the badlands of Montana in the 1960s, Yale paleontologist John H. Ostrom (1928–2005) made a discovery that would forever change our perception of dinosaurs. Ostrom named his new dinosaur *Deinonychus*.

Before Ostrom's find, it was widely accepted that dinosaurs were cold-blooded, dim-witted beasts. But the features of *Deinonychus*—especially the sickle-like killing claw found on each foot—suggested not a slow, stupid animal but an agile predator.

This single find re-energized the field of paleontology. Known as the Dinosaur Renaissance, the resulting research ultimately led to the view of dinosaurs we hold today—that they were active creatures with sophisticated behavior.

Today Museum scientists continue conducting expeditions across the globle, furthering our understanding of Earth and its life, and continuing the Peabody Museum's longstanding tradition as a leader in scientific research and education.

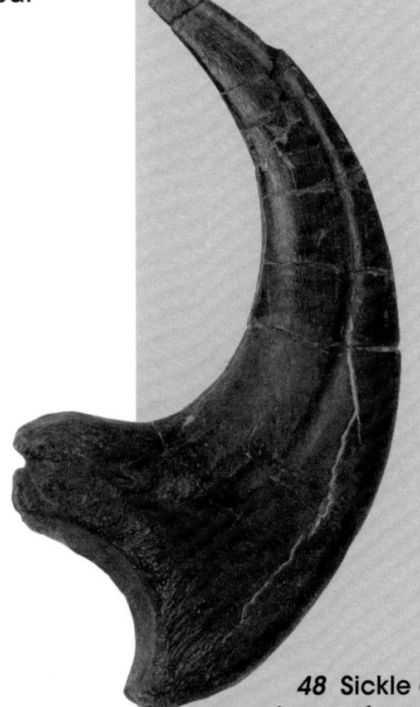

48 **Sickle claw bone of raptor dinosaur *Deinonychus antirrhopus* Carbon County, Montana (shown approximately actual size)**

49 **Illustration of *Deinonychus antirrhopus* that accompanied Ostrom's original description (1969) Artist, Robert Bakker (1945-)**

From the tropics to the poles, Yale Peabody Museum researchers have, do, and will continue traveling the globe to study our planet Earth.

Today, the Museum is home to 10 scientific divisions:

Anthropology
Botany
Entomology
Historical Scientific Instruments
Invertebrate Paleontology
Invertebrate Zoology
Mineralogy and Meteoritics
Paleobotany
Vertebrate Paleontology
Vertebrate Zoology

With each new discovery—and with each new object—we learn more about our world.

With topics of species loss and climate change dominating headlines, it is the responsibility of institutions like the Peabody Museum to work toward a better understanding of our world, our place within it, and our future role as both residents and protectors.

Opposite top

50 **Yale students collect insects attracted to a building by a mercury vapor light for a laboratory class at the Archbold Biological Station in Venus, Florida.**

Opposite bottom left

51 **Team members load vehicles in the Wadi el-Hôl, where early alphabetic inscriptions were found by curator John C. Darnell and the Yale Egyptological Institute in Egypt.**

52 **Curator Jay Ague explains the geology of Connecticut to a teacher training institute class on a field trip to Hammonasset State Park.**

Opposite bottom right

53 **The DSV *Alvin* being lowered from the R/V *Atlantis* at Manning Seamount in the North Atlantic.**

54 **Yale undergraduates Gwen Antell (seated) and Sara Kahanamoku-Snelling (recording) measure a stratigraphic section in Barbados.**

Above

55 **Peabody ornithologists found the first record of the Burrowing Owl in Suriname, which represents a significant range extension of the species in South America.**

Left

56 **Preparator Marilyn Fox works on the delicate skull of a Triassic fossil crocodile relative.**

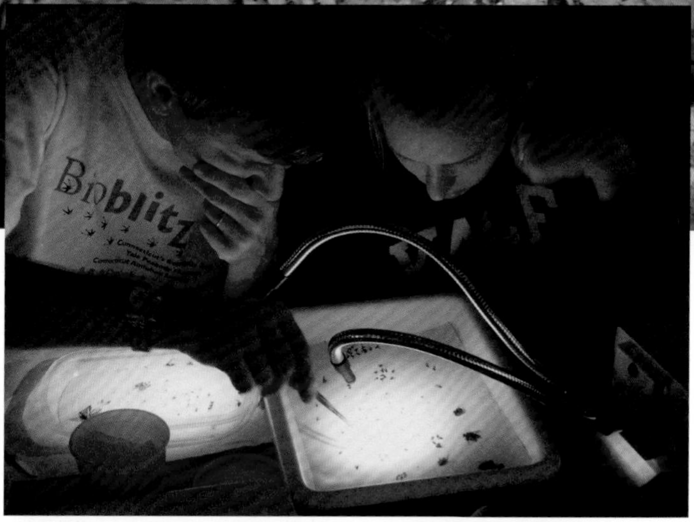

Above

61 Botany curator Michael Donoghue with Deren Eaton, Ivalú Cacho, and a parataxonomist collecting *Viburnum* in Chiapas, Mexico.

Left

62 Entomologist Larry Gall and Yale graduate student Nicole Palffy-Muhoray sort sweep-net samples of insects during the 24-hour Peabody BioBlitz in Stratford, Connecticut.

63 Yale paleontology graduate student Holger Petermann prepares to make a sturdy plaster jacket around a fossil in the Petrified Forest National Park, Arizona.

Opposite left

57 Yale students collect aquatic insects at the Yale Camp at Great Mountain Forest in the Litchfield Hills of northwest Connecticut.

Opposite right, from top

58 Undergraduate Melina Delgado catalogues *Cycadeoidea* trunks.

59 Volunteer Danica Meier and student employee Max Lambert work on pythons to be prepared as skeletons, skins, and tissue samples.

60 Paleobotanists Peter Crane and Shusheng Hu, and students Emma Locatelli and Maya Midzik, collect Cretaceous lignite on Block Island.

YPM 14080

2

Discovering Nature

Among the Peabody's artifacts and specimens are those that document the discovery of nature.

Collected from across the globe, these specimens are the first of their kind to be scientifically described. Or, they provide new insights that lead to a deeper understanding of our planet and its life.

But just as the discovery of new species can deepen our knowledge of life's diversity, the development of new technology allows us to see the world in new ways.

Presented here are just a few of the notable "firsts" housed within the collections of the Peabody Museum.

Opposite

64 Cercopithecus lomamiensis **is only the second new species of African monkey described during the past 30 years. Named in 2012 by Peabody curator Eric Sargis and a worldwide team of researchers, it was well known to local people when "discovered" by the scientists in 2007.**

Natural history involves not only naming new forms of life but also identifying and studying behavior. In 1903 Russian physiologist Ivan Pavlov (1849–1936) presented his work on "conditional reflex."

Pavlov and his assistant discovered this concept by famously examining salivation by dogs during feeding. After receiving food accompanied by a specific sound for some period of time, a dog would salivate at just the presentation of the sound alone.

65 **Saliometer (drool collector) owned by I.P. Pavlov**
ca. 1916

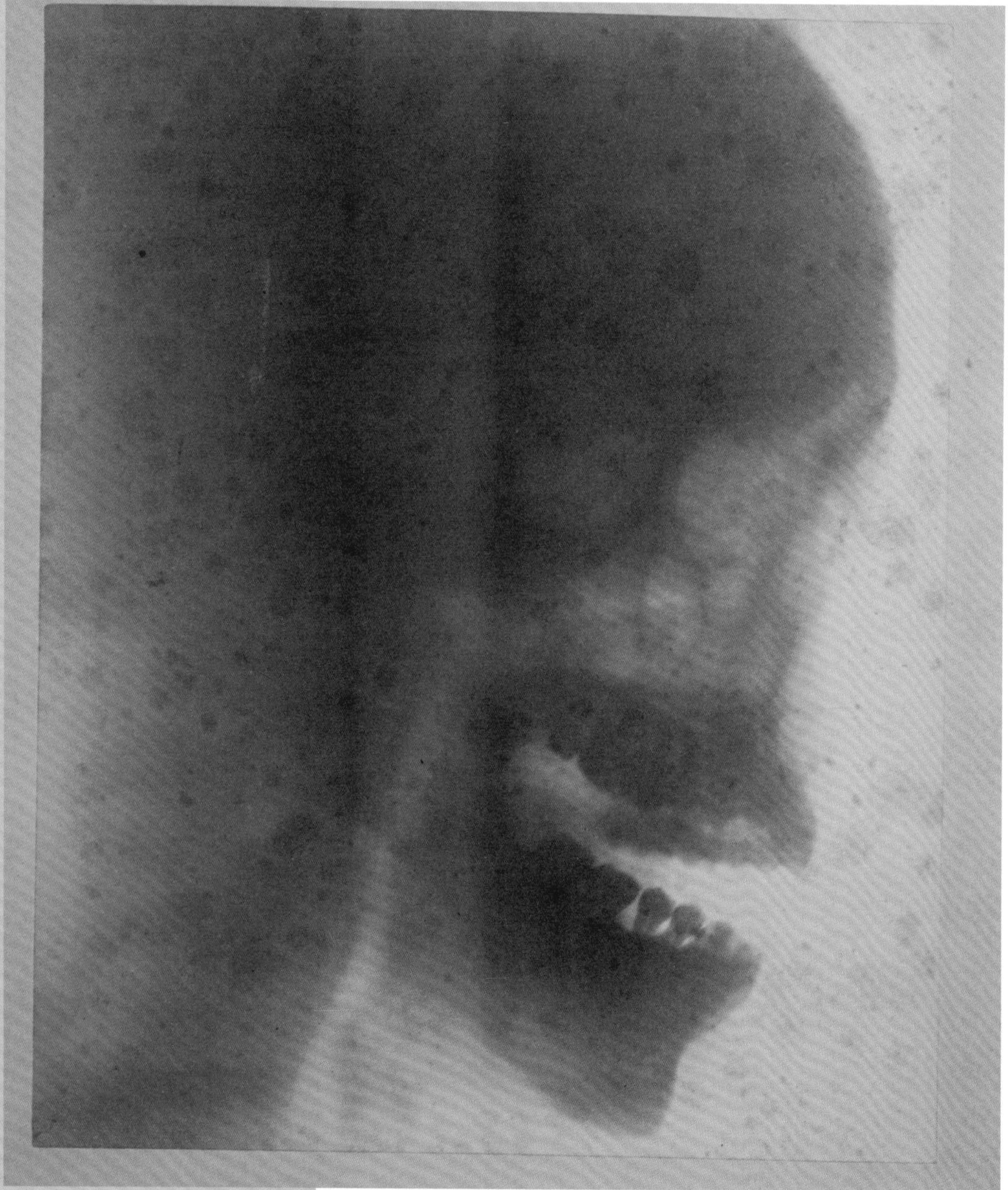

This radiograph is from a set created by Arthur W. Goodspeed (1860–1943) of the University of Pennsylvania.

Goodspeed unintentionally created the world's first X-ray image in 1890. Five years later, German physicist Wilhelm Röntgen's announcement of his similar discovery prompted Goodspeed to revisit the process that led to his first image.

This set of X-rays is likely among the earliest surviving radiographs in the United States, setting the stage for an entirely new way to study the world.

66 **Radiograph of a human head**
1896

67 Fagopsis longifolia **is an extinct species related to the beech family. First described in 1909, today it is one of the most common plants known from Colorado's Florissant Fossil Beds National Monument.**

Fagopsis longifolia
Eocene Epoch
(34 million years ago)
Teller County, Colorado
(type specimen)

Discovering New Species

Scientists have identified nearly two million species of animals, plants, and other forms of life on Earth. Each species is defined by a type specimen—the single specimen on which the species is based.

The Peabody is home to more than 100,000 type specimens. They are an invaluable catalog of Earth's life.

And exploration continues: scientists estimate that most biological species remain undiscovered.

68 **Named in 2014, this new species of leopard frog was discovered on New York City's Staten Island—one of the most densely populated areas of the world. It is one of the very few new species of vertebrates discovered in the northeastern United States in the last century.**

Leopard frog
Rana kauffeldi
Richmond County, New York
(type specimen)

69 **Not all "type specimens" are biological. Type specimens are also used to document geological diversity. This new mineral species was named in 1878 in honor of the location of its discovery.**

Fairfieldite
Fairfield County, Connecticut
(type specimen)

Marsh's Prehistoric Parade

O.C. Marsh brought the discovery and naming of dinosaurs to an unprecedented scale. His collecting efforts between the 1870s and 1890s resulted in literally tons of fossils transported from the American West to New Haven.

From these fossils Marsh named more than 500 new species, including those shown here. Many, such as *Brontosaurus*, *Stegosaurus*, and *Triceratops*, became household names.

Marsh was a key figure in bringing an understanding of prehistoric Earth to the public consciousness.

_70_ Sauropod dinosaur vertebra
Apatosaurus ajax
Marsh, 1877
Jurassic Period
(150 million years ago)
Jefferson County, Colorado
(type specimen)

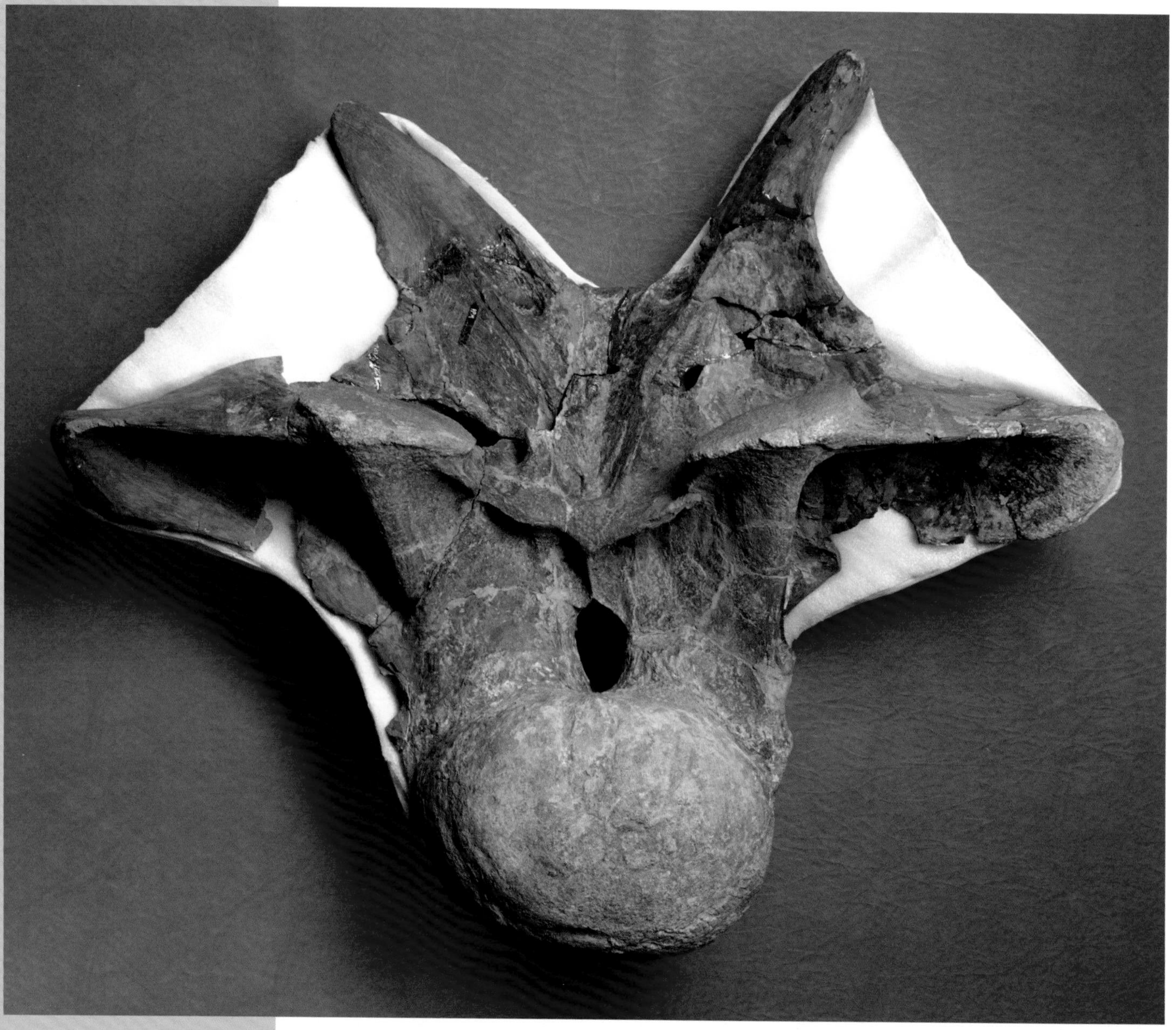

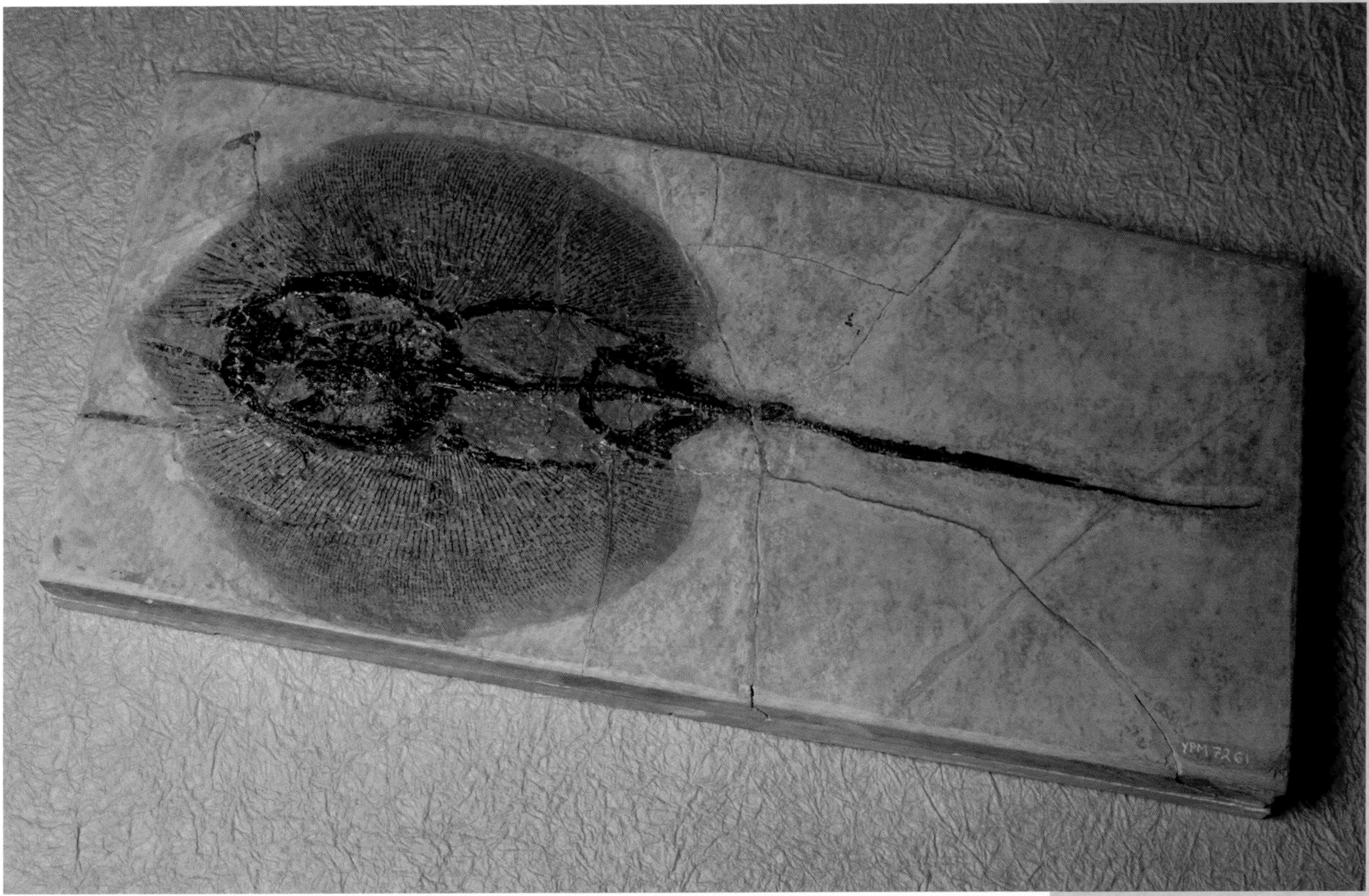

71 Freshwater ray
Heliobatis radians
Marsh, 1877
Eocene Epoch
(50 million years ago)
Lincoln County, Wyoming

72 Early mammal jaw
Docodon striatus
Marsh, 1881
Jurassic Period
(150 million years ago)
Albany County, Wyoming
(type specimen)

73 **This early model is the first life-size model of a giant squid. Shown here on display, it was constructed for the Peabody Museum by A. E. Verrill and J. H. Emerton, and was based on examples of the giant squid that washed ashore in Newfoundland.**

From Myth to Museum: Describing the Kraken

The discovery of the giant squid *Architeuthis* dates to Greek philosopher Aristotle, who first described a large squid he named *teuthus* around 350 BCE. Dutch biologist Jappetus Steenstrup scientifically named the squid in 1857.

Modern scientific study began with Yale's first professor of zoology Addison E. Verrill. Studying specimens that washed ashore along the coast of Newfoundland in the 1870s and 1880s, Verrill was able to provide new insights into the animal's anatomy.

These and other specimens were used to reconstruct the life-size model that terrorizes the Peabody's lobby today..

74 **Giant squid beak and tentacle**
Architeuthis dux
Newfoundland, Canada

A long-standing scientific mystery, *Opabinia* was a curious creature with five eyes and a long proboscis that ended in a "toothed" claw. Research by Yale's Derek Briggs has advanced our understanding of it and other animals from the Earth's Cambrian Period, placing them as early members of known animal groups. *Opabinia* is now recognized as a primitive arthropod.

75 **Primitive arthropod**
Opabinia regalis
Cambrian Period
(505 million years ago)
British Columbia, Canada

Opposite

76 **The Peabody's life-size 37.5-foot (11.4-meter) model of the giant squid *Architeuthis dux*. It was made in the 1960s by Henry Townshend, Edward Migdalski, and preparators George Rennie, Ralph C. Morrill, and Rollin Bauer.**

77 ***Aegirocassis*** **as it might have appeared in life.**

Evolutionary Oddities

The science of paleontology is often based on incomplete fossils; complete skeletons are extremely rare. But even more challenging are the fossils of those animals that are so unique—so bizarre—they defy comparison to anything living today. Like all questions in science, continued research and new discoveries provide new insight, helping scientists solve the mysteries of these ancient beasts.

This newly discovered fossil is from an extinct group of arthropods (the same group that includes insects, spiders, and crustaceans) called anomalocaridids.

Most anomalocaridids were large active predators. But seven-foot-long *Aegirocassis*—named in 2015 by a team of researchers that included Peabody curator Derek Briggs—was instead a giant filter-feeder, cruising the waters to capture tiny, nutritious plankton.

78 **Anomalocaridid arthropod**
Aegirocassis benmoulai
Ordovician Period
(480 million years ago)
Morocco

Look closely. Collected in 1990, this 570-million-year-old rock contains the impressions of some of Earth's first multicellular life.

Fossils from this time reveal a suite of frond, disc, and worm-like organisms, but their relationship to modern animals is controversial. In 1992 Yale adjunct professor Adolf Seilacher (1925–2014) proposed an entirely new kingdom in which to classify them: the Vendobionta.

79 **Vendiobionta assemblage (cast)**
Precambrian
(570 million years ago)
Newfoundland, Canada

Peabody's Primeval Forest

With more than 1,000 specimens in its collection, the Peabody is home to the largest collection of cycadeoids in the world. Cycadeoids are an extinct group of cycad-like plants that thrived during the Age of Dinosaurs.

It was the interest of the Museum's first paleobotanist George R. Wieland in these enigmatic plants that led to this collection. Through its study he published his very influential research in the early 1900s.

Wieland also created a major collection of thin sections that exposed the internal details of cycadeoid anatomy, including that of their flower-like reproductive structures.

80 **George Wieland stands amid the exhibits in the Peabody Museum's Great Hall in November 1931. Displayed in the foreground are several clusters of 125-million-year-old cycadeoid trunks (*Cycadeoidea marshiana*).**

Travel and Technology

From the quarries of New York State to the islands of the Indian Ocean, Peabody expeditions have revealed both large and small. Critical to the advancement of science is also the development of new technologies that provide new ways of looking at—and interpreting—the natural world.

A small quarry in Central New York has produced one of the world's finest examples of fossil preservation: Beecher's Trilobite Bed. From this thin layer of shale, the fossils reveal not only the hard outer skeleton of trilobites but also their delicate limbs, gills, walking legs, and even antennae!

This site—named after Yale geologist Charles E. Beecher (1856–1904), who heavily excavated there in the 1890s—provided scientists with an unparalleled view of these ancient sea-goers.

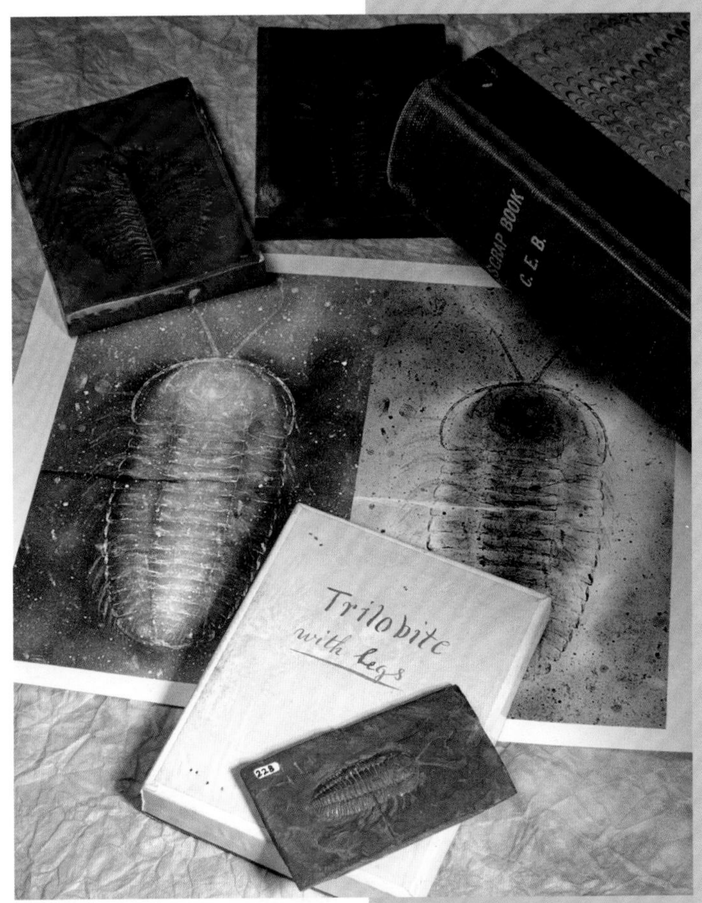

Above

81 **The 450-million-year-old trilobite *Triarthrus eatoni* with a grouping of Beecher's research materials, including a photograph of X-rays of the specimen that shows details of gills, limbs, and antennae. A metal mold is at the top left.**

Below

82 **Charles E. Beecher (seated, right) at the Oneida County fossil excavation site in upstate New York.**

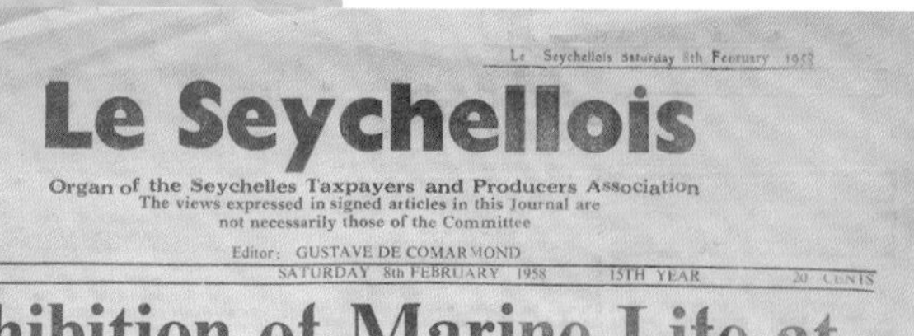

Le Seychellois Saturday 8th February 1958

Le Seychellois

Organ of the Seychelles Taxpayers and Producers Association
The views expressed in signed articles in this Journal are not necessarily those of the Committee

Editor: GUSTAVE DE COMARMOND

NO. 681 SATURDAY 8th FEBRUARY 1958 15TH YEAR 20 CENTS

Exhibition of Marine Life at the Seychelles College

Two American Scientists (Drs Alan Kohn and Willard Hartman) explained their specimens and gave talks on their expedition.

Mr John H. Hayward, former Warden, K… beautiful film of East African ani…

A most successful exhibition of marine life presented by Dr Alan Kohn and Dr Williard Hartman was opened at the Biology Laboratory of the Seychelles College on Monday 3rd February at 5.30 p.m.

There was a large number of people present at this Exhibition and the two scientists showed them round the various exhibits laid on the tables of the Laboratory.

The exhibits comprised the following:

TYPES OF ANIMALS LIVING IN THE SEA:

The first exhibits were SPONGES and species shown were yellow, red, olive and brown encrustations.

POLYPS & JELLY FISH: (*Coelenterates*) The exhibit was a soft coral with an unusually large polyp. Each polyp has 8 tentacles constantly waving in search of food.

BRISTLE WORMS: (*Polychaetes*). These common animals are characterised by having segmented bodies, each segment bearing a pair of appendages.

JOINTED-LEGGED ANIMALS (*Arthropods*). The exhibits on view were some hermit crabs and a true crab, examples of crustaceans.

BARNACLES: These animals are modi-

animal, the description states that the coconut crab or *cipaille* must lay its eggs in the sea and the young crabs live in the sea for some time after hatching. The adults live on land and subsist on coconuts which they are able to husk with their large claws. The specimen on view was collected at Aldabra Island.

PLANT-ANIMAL INTERRELATIONSHIPS

CORALS & ALGAE: Two specimens were on view. The brown color is caused by minute algae living in the polyps.

GIANT CLAMS FARM ALGAE (*Benitier*): Giant clams bear small lenses in their mantle edge. These concentrate light and encourage the growth of algae.

The specimen was a very small one. There are specimens ten times that size everywhere on the reefs here.

COLOUR & CONCEALMENT ON THE REEF.

This exhibit was a reef crab with a sponge on its back. A very good disguise and it is sometimes very difficult to distinguish the crab when walking on the reefs.

HOW REEF ANIMALS EARN A LIVING.

DEPOSIT FEEDERS: Sea Cucumbers

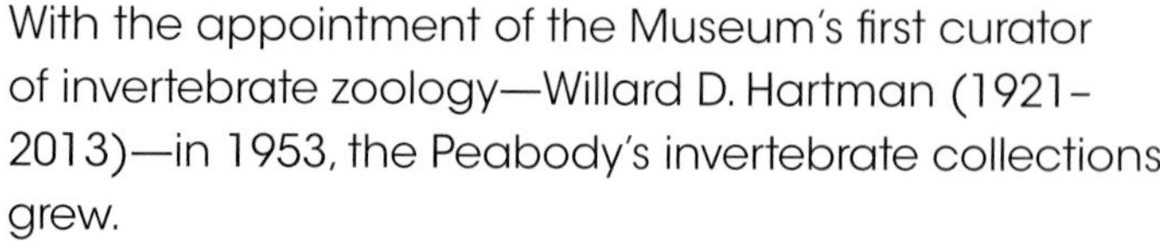

With the appointment of the Museum's first curator of invertebrate zoology—Willard D. Hartman (1921–2013)—in 1953, the Peabody's invertebrate collections grew.

Hartman collected many specimens from the tropics. In particular, the Yale Seychelles Expedition of 1957–1958 led to an increased understanding of the distribution of animals and plants on islands.

Above

83, 84 **Under Willard Hartman's stewardship the Peabody Museum's invertebrate zoology holdings expanded, particularly in areas of his principle research interest, the systematics and evolution of sponges and their association with coral reefs.**

"Exhibition of Marine Life at the Seychelles College" ***Le Seychellois*** **(1958)**

85 **Seychelles coral collected by Willard Hartman on the Yale Seychelles Expedition.**

Tubipora **sp.**
Peros Banhos Atoll, Indian Ocean

The scale bar measures 30 millimeters (about 1 inch).

Located in the Indian Ocean, the Republic of Seychelles is composed of 115 islands. Two of these are home to the rare palm *Lodoicea*. Known as the love nut, its fruit contains the largest seeds within the plant kingdom. Its seeds do not remain viable in salt water, keeping it from spreading to nearby islands.

86 **Seychelles love nut (with an acorn of the white oak for comparison)**
Lodoicea maldivica
Praslin Island,
Republic of Seychelles

A. Lakes 78

3

From the Ends of the Earth

By the late 1700s, a growing interest in the natural sciences and increasing awareness of Earth's biological and geological diversity generated efforts to carry out formal scientific expeditions.

One of the early voyages of scientific discovery carried a young naturalist named Charles Darwin. The U.S. Exploring Expedition took place not long after, providing some of the first collections for the new Smithsonian Institution—and to the Peabody Museum on its establishment.

Expeditions remain just as critical to the work of museum scientists today.

87 **Pioneer dinosaur hunter Arthur Lakes was employed by the Peabody's O.C. Marsh to excavate dinosaur fossils in the American West. Lake's excavations would uncover fossils of *Stegosaurus, Allosaurus, Brontosaurus,* and many more. Also an artist, Lakes captured many scenes in pastel sketches and watercolors.**

Opposite

Professor Benjamin F. Mudge (on the right, one of O.C. Marsh's "bone collectors") and a helper excavate a skull near Morrison, Colorado, in 1878.

Morrison Quarry 1
Watercolor on paper
Artist, Arthur Lakes (1844–1917)

America's First Great Expedition

Pressed by President John Quincy Adams, in May 1828 Congress approved an expedition around the world: the U.S. Exploring Expedition. Just as the American West was being explored by the young country, now too would the sea.

Ten years later, the expedition would begin. Six sailing ships carried nearly 350 men, including a small team of artists and scientists. Serving as geologist was Yale's James Dwight Dana.

In total Dana collected examples of 300 fossil species, 400 speces of coral and their relatives, and 1,000 species of crustaceans.

Although Dana was tasked as the expedition's geologist, his collections and research extended beyond rocks and minerals. From a specimen collected during the expedition by the crew's naturalist Charles Pickering, Dana named a new species of crab *Cancer magister* in 1852 (now classified as *Metacarcinus*). The specimen on the opposite page is one of only a few existing crustaceans from the expedition. All others were destroyed in the Great Chicago Fire of 1871.

_88_ Collected in the South Pacific, this new species of coral was studied and named _Madrepora carduus_ (now classified as _Acropora_) by Dana in 1846, four years after the end of the expedition. Dana's lavishly illustrated volume _Zoophytes_ details many of his finds from the expedition.

Coral
Acropora carduus
Fiji Islands

Pencil sketch of
Acropora carduus
Artist, Alfred Thomas Agate
(1812–1846)

89 This line and shell hook, still with its original tag, was among the 4,000 ethnographic objects collected during the expedition.

Sinnet line
and shell hook
Samoa (Polynesia)

When the expedition pulled into port, exploration continued on land. Shown here are the fossils of an extinct group of plants called seed ferns, collected by Dana, and a fern collected by the expedition's botanists.

Found from South America to Australia, the seed fern *Glossopteris* would prove vital to early ideas of continental drift and the supercontinent Pangaea.

This modern fern is the type specimen of the species—the single specimen on which the species is based.

Above, left

90 Fossil seed fern leaves
Glossopteris browniana
Permian Period
(260 million years ago)
New South Wales, Australia

Above, far right

91 Fern
Hymenophyllum formosum
Tahiti, French Polynesia
(type specimen)

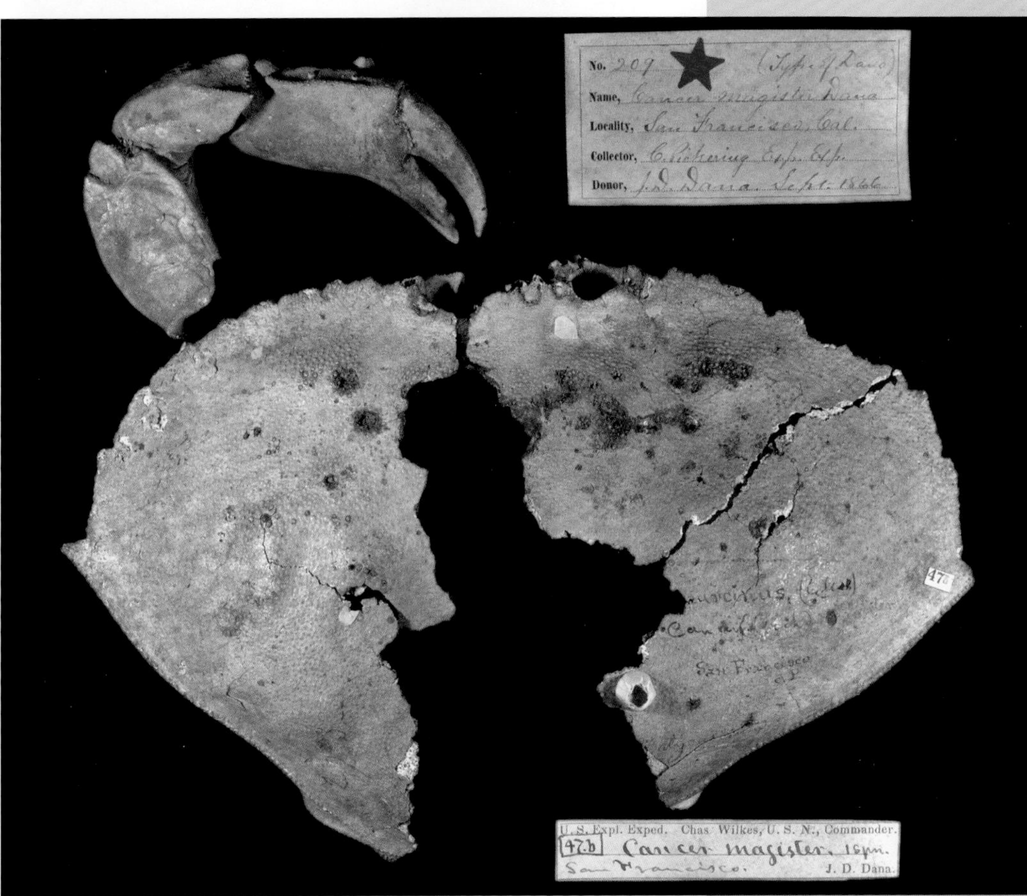

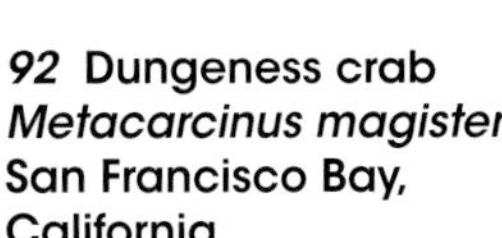

92 Dungeness crab
Metacarcinus magister
San Francisco Bay,
California

93 **Some of the early horse fossils from O.C. Marsh's "hatful of bones"**
Protohippus parvulus
Miocene Epoch
(15 million years ago)
Kimball County, Nebraska
(type specimen)

Brontosaurs, Bullets, and Buffalo Bill

America's first professor of paleontology and the Peabody's first director O.C. Marsh made his name leading expeditions to the American West.

Fueled by a passion for fossils, Marsh and his teams would ultimately discover early horses, primitive birds, and a parade of now-iconic dinosaurs.

In 1868, after a scientific meeting in Chicago, Marsh participated in an excursion along the still under-construction transcontinental railroad—his introduction to the American West.

Along the route Marsh negotiated a stop in Antelope Station, Nebraska, where "human remains" had been recently reported. Quickly inspecting the fossils, Marsh didn't see human remains but instead fragments of ancient reptiles and mammals excavated from a nearby well. He asked the station agent to collect the rest of the fossils.

When the train passed through on the return trip east, Marsh was given a "hatful of bones," some of which were from a new species of horse Marsh named *Equus parvulus* (now classified as *Protohippus parvulus*).

94 **Hand-drawn map of the U.S. Army scouting party that took O.C. Marsh and the Yale College Scientific Expedition of 1870 (route shown in red) from July 14 to 30 through the Sand Hills of Nebraska.**

95 **O.C. Marsh (standing, center) and his guides with the Yale students (seated) of the 1872 expedition.**

Marsh led his first expedition into the American West in 1870, the first of four Yale College Scientific Expeditions. From 1870 to 1873, Marsh and Yale students explored America's badlands in search of fossils.

Escorted by the U.S. Army, the team also included none other than William F. Cody—better known as Buffalo Bill—as a guide. Though Cody was called to another assignment after only one day, he and Marsh remained lifelong friends.

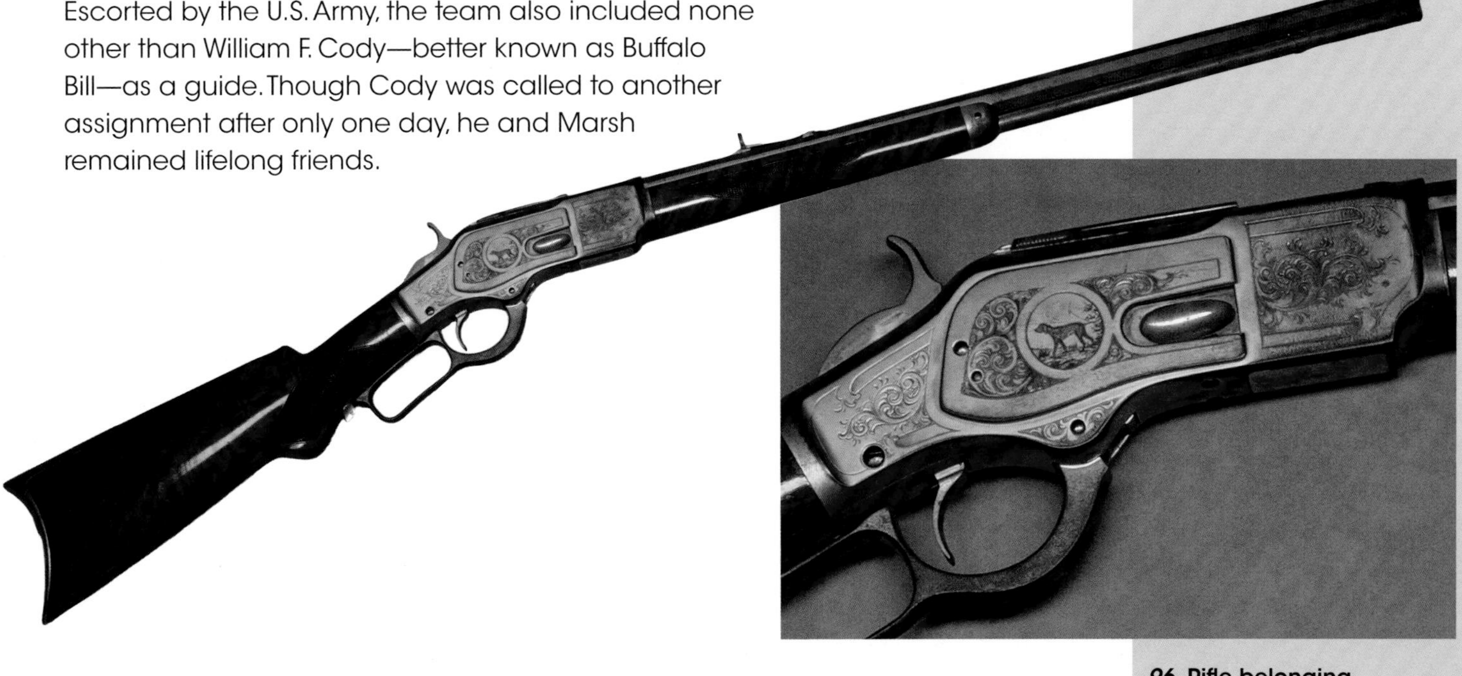

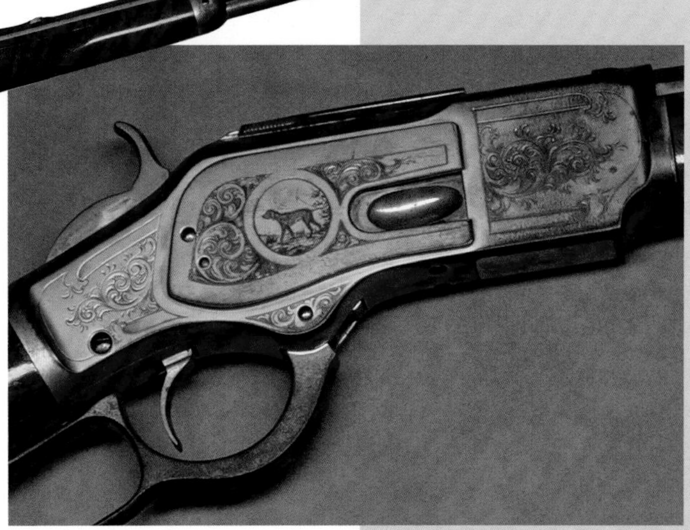

96 **Rifle belonging to Buffalo Bill.**

97 **Edward Drinker Cope**

As the American West opened and more fossils were discovered, a fierce rivalry between Marsh and Philadelphia paleontologist Edward Drinker Cope soon arose: the Bone Wars.

Marsh and Cope first met in 1864 and soon struck up a friendship. Cope even named a fossil amphibian after Marsh: *Ptyonius marshii.* In return Marsh named a fossil marine reptile *Mosasaurus copeanus.*

But Marsh's gesture was undercut by his ruthlessness. Marsh's reptile had been excavated in a quarry shown to Marsh by Cope. Unknown to Cope, Marsh made an agreement with the quarry owner to ship any new fossils to New Haven, not Philadelphia. The friendship devolved into hostility.

Marsh and Cope named over 140 new dinosaurs between them, but their bitter public feud ultimately left their reputations scarred.

98 **Bones of *Mosasaurus copeanus* Cretaceous Period (66 million years ago) Monmouth County, New Jersey (type specimen)**

99 **O.C. Marsh (with hammer) and part of his team during the 1871 expedition.**

This crated dinosaur bone was prepared and shipped to Marsh by his men in June 1892. By that time, Marsh had become the vertebrate paleontologist of the U.S. Geological Survey.

But as a result of Cope's public attacks and accusations, as well as accusations of corruption and misuse of Survey funds, Congress eliminated Marsh's position and funding, effective that August 1.

The crate arrived in New Haven 18 days later.

100 **Dinosaur bone with 1892 original crate and packing.**

101 **The 1925 logbook from the *Pawnee* and specimens from the Bingham Oceanographic Expeditions. *From left to right:* A flying fish (*Cypselurus vitropinna,* type specimen) and a lobster (*Nephrops binghami*) from the Caribbean Sea next to two jars of red sargassum; a velvet whalefish (*Barbourisia rufa,* type specimen) from the Gulf of Mexico, and a dragonfish (*Echiostoma ctenobarba,* type specimen) from the Caribbean Sea. A jar of green sargassum and water samples are at the far right.**

A Man and His Yachts

From 1925 to 1927, Yale graduate Harry Payne Bingham (1887–1955) led three oceanographic expeditions for his own private research, focusing on the southern Atlantic, Caribbean, and the Gulf of California. He collected thousands of specimens.

In 1959 specimens from the Bingham Oceanographic Expeditions—at Yale since the late 1920s—were incorporated into the Peabody's collections.

For his travels, Bingham used his personal yachts: *Pawnee* for the first in 1925, and a second *Pawnee* for those in 1926 and 1927. This logbook is from the *Pawnee* and Bingham's 1925 exploration.

Bingham's expeditions led to the discovery of many new species of fishes, as well as a parade of crustaceans, sea stars, and other sea life.

The seaweed (*sargassum*) shown here was collected later, in the 1930s, during an expedition by Yale and the Wood's Hole Oceanographic Institute—under the auspices of the Bingham Oceanographic Collection.

From the Darkest Depths

Stretching more than 600 miles (1,000 km), the New England Seamount Chain is a series of more than 20 extinct volcanic peaks in the deep waters of the North Atlantic.

Whereas early studies focused mainly on the geology of the chain, recent expeditions have produced the first biological collections from the area. These specimens, many of which are now housed at the Peabody, reveal a fascinating deep-water fauna.

Collected from Bear Seamount in 2004, the most western peak within the chain, these specimens reveal an amazing diversity of life within the deep Atlantic, thousands of feet below the surface. More than 400 species have been identified from these expeditions.

_102_ Deep sea animals from the North Atlantic Bear Seamount. _Left to right:_ Lovely hatchetfish (_Argyropelecus aculeatus_), treadfin dragonfish (_Echiostoma barbatum_), barbeled dragonfish (_Melanostomias bartonbeani_), a delicate white coral (_Chrysogorgia tricaulis_), and a brittle star (_Asteroschema_ sp.) wrapped around a red coral (_Paragorgia johnsoni_).

PROVENANCE ENTRY DATE EXCAVATOR'S NO.
AE I.5
66-11-38
1991
ARMINNA EAST, 25 FEB '62
TOMB I
TOSHKA WEST, FEB '62
TOSHKA
CEM. D,
TOSHKA WEST,
CEM. C, TOMB
WEST, 26 FEB '62 AW M
DEIR AMBINIRA
Room 8
TOSHKA WEST, 14 FEB '62 TW D 5.4
TW D 5.5
CEM. D, TOMB
5
Tom Lovejoy
no weight
cere yellow orange; bill dark blue grey
legs and feet yellow

Saving History

Completed in 1902, Egypt's Aswan Dam was constructed to control and regulate the Nile River. To further protect from floods and droughts, as well as generate electricity, construction of a new, larger dam—able to hold five times as much water as Hoover Dam—was slated to begin in 1960.

Archaeologists soon raised concerns that the new dam would flood historical sites located upriver.

In collaboration with the University of Pennsylvania, the Peabody's William Kelly Simpson (1928–) led expeditions to excavate these sites before their destruction.

The Pennsylvania-Yale expeditions occurred from 1961 to 1963. This field register, belonging to Simpson (kneeling in this 1960 excavation staff photograph), documents objects recovered during the 1962 work.

These objects were among those excavated in 1962 from the regions of Toshka and Arminna and preserved from the flooding.

In addition to archaeological remains, the expeditions also collected these and other specimens of modern animals.

Opposite

103 **Some of the artifacts recorded in the 1962 ledger.** ***From top to bottom:*** **Kohl jar, Middle Kingdom (2030–1650 BCE), from Nubia, Egypt; a bead with hieroglyphs of Pharaoh Antef V and goddess Tawaret, 2nd Intermediate Period (1650–1550 BCE) from Toshka West, Nubia, Egypt; and a Wedjat eye amulet, Meroitic (270 BCE–350 CE) from Arminna West, Nubia, Egypt.**

The animal specimens collected included the Egyptian fruit bat (*Rousettus aegyptiacus aegyptiacus*), the Nubian spitting cobra (*Naja nubiae,* type specimen), and the common kestrel (*Falco tinnunculus rupicolaeformis*).

Our Own Backyard

Throughout the Peabody's long history, Museum scientists have traveled the world in search of its life and history. But from dinosaurs to meteorites, the natural history of Connecticut is just as vast and rich.

Many objects within the Peabody's collection were recovered from Connecticut and adjacent states, documenting New England's past and present.

Dinosaur footprints are relatively common in Connecticut, but skeletons are extremely rare. Collected in 1891, this skeleton of a dog-sized dinosaur is a true treasure of the Peabody's collection.

Once called *Yaleosaurus* in honor of Yale, this skeleton is now classified as *Anchisaurus*.

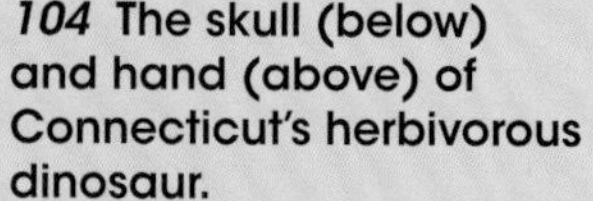

104 **The skull (below) and hand (above) of Connecticut's herbivorous dinosaur.**

Anchisaurus polyzelus
Jurassic Period
(200 million years ago)
Hartford County, Connecticut

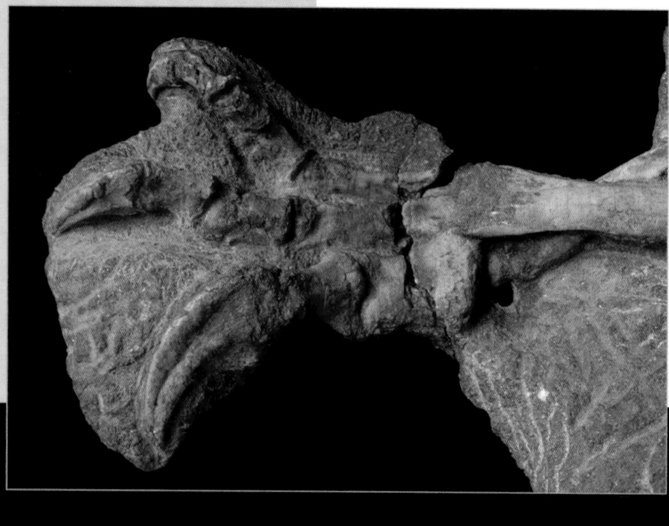

The Yale University Herbarium—located within the Peabody's Division of Botany—is home to more than 350,000 plant, fungal, and lichen specimens collected from across the world.

Among its holdings is the one of the earliest known collection of plants from Connecticut: nearly 700 botanical specimens collected in the New Haven area in 1822. Horatio Nelson Fenn (1798–1871), then a student at the Yale Medical College, created the collection as part of his studies.

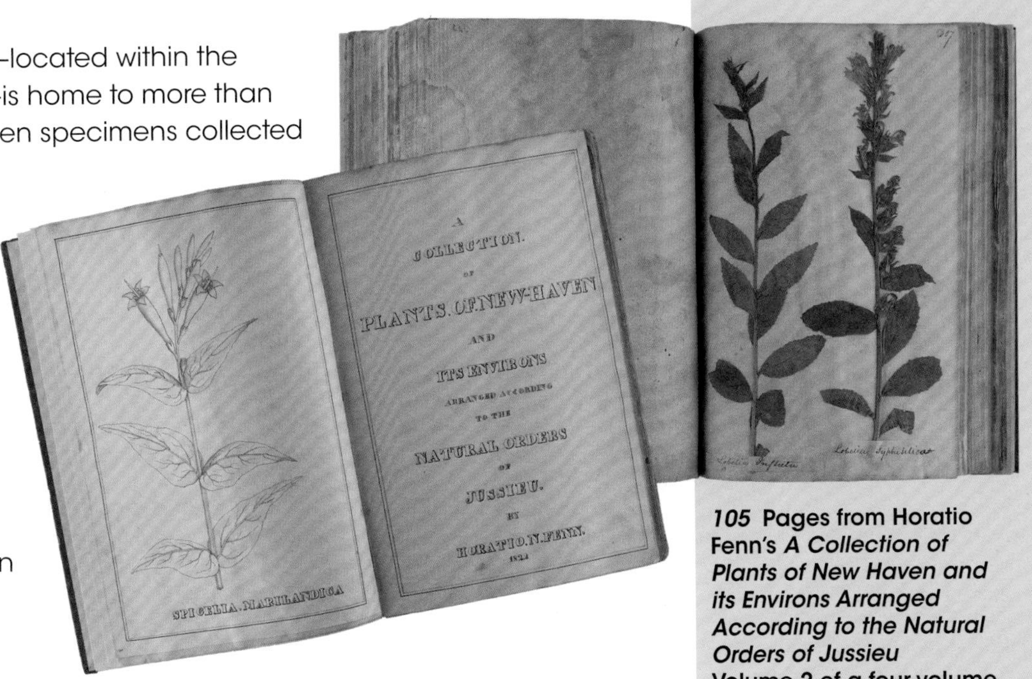

105 **Pages from Horatio Fenn's *A Collection of Plants of New Haven and its Environs Arranged According to the Natural Orders of Jussieu* Volume 2 of a four-volume set (1822)**

Over the past 200 years, five meteorite falls have been recorded in Connecticut, in Weston (1807), Wethersfield (1971 and 1982), Stratford (1974), and Wolcott (2013).

This is a large fragment of the Weston meteorite from the December 14, 1807 fall, and a small piece of the Wolcott meteorite—whose center dates back to 4.56 billion years ago, and whose crust was formed on April 19, 2013.

106 **The Weston meteorite and the smaller Wolcott meteorite (type specimen), with a copy of Benjamin Silliman's 1810 "An Account of the Meteor," from the *Memoirs of the Connecticut Academy of Arts and Sciences.***

4

Conserving Nature

As topics of habitat loss and species extinction face us today, preserving and protecting Earth's natural resources is a key mission of researchers far and wide.

From its beginnings, the Peabody Museum has played—and continues to play—a primary role in shaping conservation movements across the world.

Beginning in America and spreading beyond our shores, the influence of these efforts continue to shape theory and policy, helping preserve our home for generations to come.

Opposite

107 **Army officers at Yellowstone with buffalo heads confiscated from poacher Edgar Howell in 1894, the incident that led to George Bird Grinnell's story in *Forest and Stream* that called for protection of the park's wildlife.**

108 **Once owned by O.C. Marsh, this painting by Hudson River School artist Thomas Whittredge depicts a landscape of the type Marsh investigated during his exploration of the American West. With the opening of the West, students such as George B. Grinnell were inspired to preserve and protect these lands.**

Junction of the Platte Rivers **(1866)**
Oil on canvas
Artist, Thomas Worthington Whittredge (1820–1910)

American Conservation Is Born

At the time of the Louisiana Purchase in 1803, the American West was viewed as a vast, unconquerable wilderness. Less than a lifetime later, the West was a landscape under rapid transition that would soon lose its frontier status. Dawning recognition of the vulnerability of these inspiring landscapes led to an explosion of interest in conservation.

Among the students participating in the Yale College Scientific Expedition of 1870 was George Bird Grinnell (1849–1938). Led by the Museum's first director O.C. Marsh—and guided by scout "Buffalo Bill" Cody—the expedition secured Marsh's position as America's foremost collector of fossils.

But for Grinnell, the exploration of the American West left a profound impression, which would later define him as one of the first leaders of America's conservation movement.

109 **Taken in Chicago, this team photograph of the 1870 Yale College Scientific Expedition with O.C. Marsh (standing, center) includes George Bird Grinnell (standing, second from left).**

Grinnell's love of birds developed at an early age, likely inspired by his schooling at the estate of John James Audubon, where he was taught by Audubon's widow Lucy.

His interests in birds and the natural world eventually led him to study at Yale. Here, his graduate research focused on *Geococcyx californianus*, the greater roadrunner.

110 **Greater roadrunner**
Geococcyx californianus
Los Angeles County, California

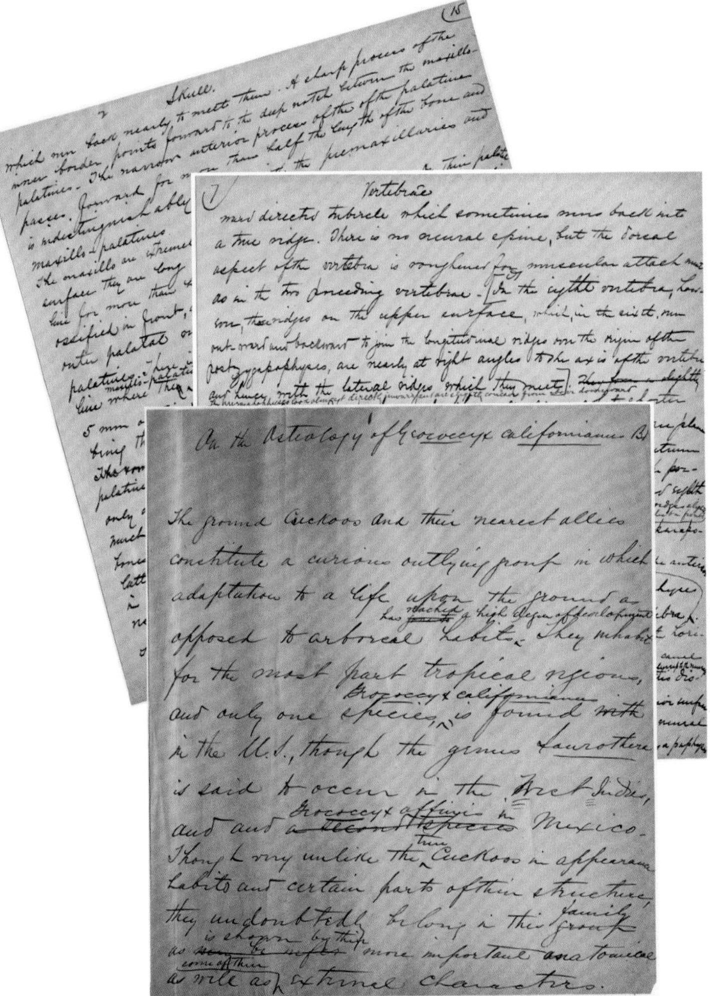

On the Osteology of Geococcyx californianus

The ground Cuckoos and their nearest allies constitute a curious outlying group in which adaptation to a life upon the ground as opposed to arboreal habits has reached a high degree of development. They inhabit for the most part tropical regions, and only one species, Geococcyx californianus, is found within the U.S., though the genus Saurothera is said to occur in the West Indies, and a Geococcyx affinis in Mexico. Though very unlike the true Cuckoos in appearance, habits and certain parts of their structure, they undoubtedly belong in this family as is shown by some of their more important anatomical as well as external characters.

111 **George B. Grinnell's handwritten dissertation draft, "On the Osteology of *Geococcyx californianus*," select pages**

In 1872 Yellowstone was established as the first national park. But without appropriate funds for protection, rock formations were destroyed by visitors and wildlife slaughtered.

112 **George Bird Grinnell in 1925 standing on the glacier named for him.**

With his appointment as editor of *Forest and Stream* magazine in 1876, Grinnell had an enviable pulpit from which to spread his conservation message. After a bison poaching incident in 1894, Grinnell published an exposé in *Forest and Stream*, leading to public outcry.

Later that year President Grover Cleveland signed the Act to Protect the Birds and Animals in Yellowstone National Park.

Grinnell and others also pushed for the creation of a national park in northern Montana and were rewarded when President William Howard Taft in 1910 signed the bill establishing Glacier National Park as the country's 10th national park.

Today, of the park's 17 glaciers documented since 1997, 13 have noticeably receded, some to just one-third of their estimated historical maximum size. Only 26 named glaciers still exist of the 150 glaciers present in 1850, mere remnants of their previous size.

Below

113 **Grinnell Glacier in 1887 (left) and in 2013 (right) as documented by the U.S. Geological Survey's Repeat Photography Project.**

To further his message of nature and wildlife conservation, Grinnell also founded the Audubon Society of New York—the precursor of today's National Audubon Society.

For his lifelong devotion to conservation, George Grinnell is generally regarded as the "father of American conservation." In 1925 he was awarded the Roosevelt Memorial Association Distinguished Service Medal for these efforts. Earlier, in 1897, Teddy Roosevelt and Grinnell had co-founded the Boone and Crockett Club.

Vol. I. February, 1887. No. 1.

The Audubon Magazine

J. J. Audubon

Published in the Interests of The
AUDUBON SOCIETY
for the
PROTECTION OF BIRDS

FOREST AND STREAM PUBLISHING COMPANY.
NEW YORK.

Annual Subscription, 50 Cts. Single Copy, 6 Cts.

***114 Audubon* magazine (reprint)**
Volume 1, Number 1 (1887)

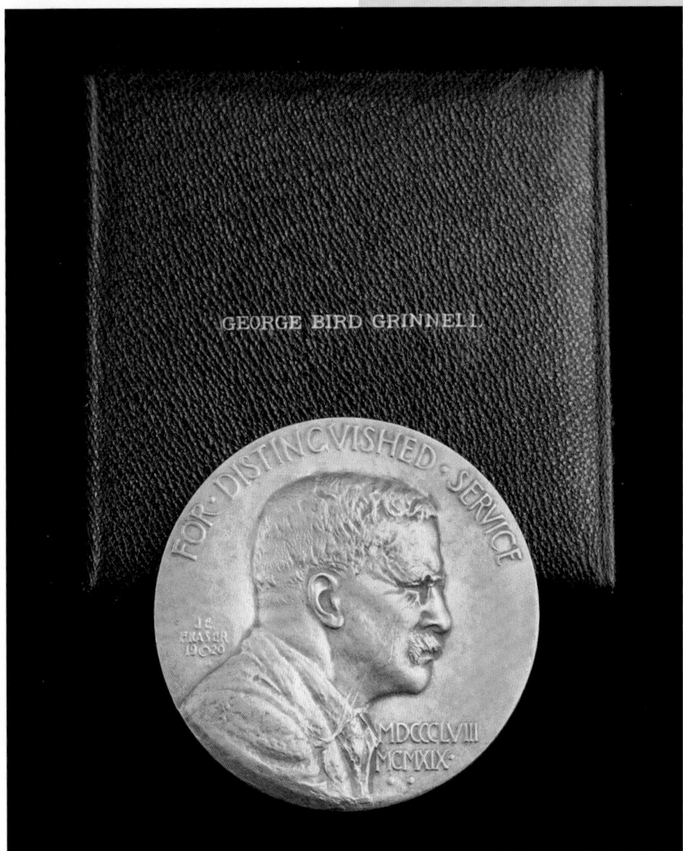

***115* Roosevelt Memorial Association Distinguished Service Medal, presented to George Bird Grinnell, 1925**

FIRST ANNUAL MEETING OF THE XERCES SOCIETY

PROGRAM

Announcing . . .

THE XERCES SOCIETY

To everyone who wants to help perpetuate rich, natural populations of butterflies:

Butterflies are declining. They are subject to the same environmental pressures faced by peregrines and pumas — but until now there has been no butterfly defense movement in North America. The Xerces Society has been founded to resist the destruction of butterfly populations. We channel the energies of the people to protect endangered butterflies, politically and otherwise. The Xerces Institute promotes and will eventually sponsor research in the ecology and habitat needs of butterflies.

HABITAT PROTECTION IS THE MAIN FRONT. The Xerces Blue, formerley of the Bay Area, was the first American butterfly to be lost.

NO FURTHER EXTERMIN... SHOULD BE TOLERATED.

Symbol of the Xerces Society

Glaucopsyche lygdamus (Dbldy.)
ENT.447194 (n=7)

dedicated to the preservation of butterflies and their habitats

Evolution of a Movement

As conservation has evolved and become a global concern, Peabody scientists have continued to provide critical leadership.

Charles L. Remington (1922–2007) was at the forefront of raising awareness that the human population had become a planetary force. For much of the latter part of the 20th century, humans were regarded primarily as threats to biodiversity.

Joining Yale in 1948, Remington was the Peabody's first curator of entomology. His love of insects guided his research, but he was also a leader in the conservation movement of the 20th century.

Among his many efforts, Remington co-founded with Paul Ehrlich Zero Population Growth in 1968, a society dedicated to issues of human overpopulation and environmental sustainability.

Remington's focus on both local and global scales served as a model for 20th century conservation efforts.

Driven by his fascination for butterflies and other insects, Remington was instrumental to the founding of the Xerces Society in 1971—the first conservation society focused on invertebrates. The society's name is based on the first American butterfly species to become extinct due to human activities.

116 **Taken on the first manned mission to orbit the Moon—Apollo 8—this image became an icon for the conservation movement. For the first time, Earth could be viewed hanging in space, solitary and fragile.**

Opposite

117 **The program for the 1974 annual meeting of the Xerces Society and a postcard announcing the founding of the Society, along with specimens of Xerces blue butterflies.**

Glaucopsyche lygdamus xerces
San Francisco County, California

At the Peabody Museum, Remington assembled one of the nation's great insect collections. Among these specimens is one of the world's largest assemblages of gynandromorphs—animals with bodies displaying both male and female features.

In some instances the features of both sexes are found across the body. But in others they are clearly divided: one side female, the other male.

118 **Charles Remington**

119, 120 **Examples of gynandromorph butterflies.**

At left: Doxocopa laurentia **from Brazil.**

Below: Lycaena gorgon **from California.**

Remington was also widely known for his popularization of *Magicicada*, the 17-year cicadas. He established a successful preserve for the insects in Hamden, Connecticut—the first of its kind and chosen based on the Peabody's continuous records from the New Haven area dating back to 1843.

121 **Cicada collection from broods dating from (counterclockwise from bottom left) 1843, 1860, and 1877. The map shows sites surveyed for broods (open circles) and locations where cicadas were present (solid circles).**

Magicicada septendecim **New Haven County, Connecticut**

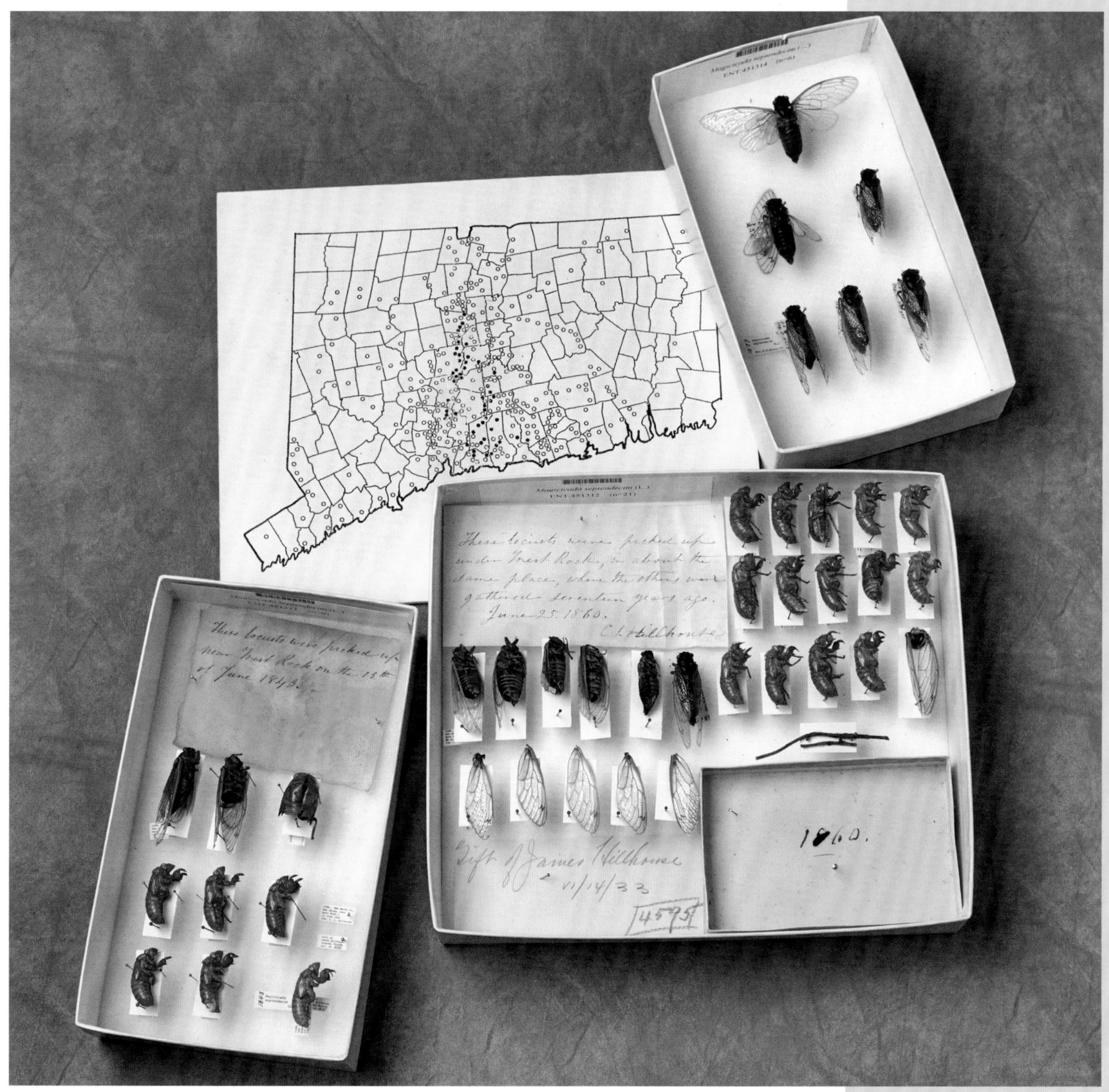

Conservation Today and Beyond

As the movement to limit human population growth was not having the desired effect, some scientists recognized that the fate of biodiversity would depend ultimately on local communities. They saw people who depended on biologically rich ecosystems as allies.

Former Peabody director Alison Richard has been a pioneer in the development of such community-based conservation approaches. The Bezà Mahafaly Reserve, where Richard has worked for four decades, is the site of a remarkable and successful partnership between community members and researchers, focused on common goals of ecosystem health.

Community-based conservation depends on supporting livelihoods through means that avoid compromising ecosystems. At Bezà, the production of salt from deposits left by an ancient sea allows creation of a desirable export product with minimal harm.

122 **The sifaka (*Propithecus verreauxi*) has been the subject of research at Bezà since 1974. The Peabody hosts a long-term database covering more than 25 years of observations collected from over 750 marked animals.**

123 Monitoring data allows community members to evaluate the ongoing health of the ecosystem. Here, Efiteria, who carries out radiated tortoise (*Astrochelys radiata*) monitoring, holds one of his study subjects as professors Joel Ratsirarson and Alison Richard (left) look on.

124 At Bezà, research and conservation action go hand in hand. Here, Joel Ratsirarson and Alison Richard confer over lemur census data while other team members look on.

5

Evolving Science

Interest in the relationship between birds and dinosaurs goes back to the very early history of the Peabody: O.C. Marsh was among the first paleontologists to recognize affinities between living birds and extinct dinosaurs. Today, scientists overwhelmingly agree: birds are living dinosaurs.

But for much of the 20th century this idea was dismissed, until another Peabody paleontologist—John Ostrom—resurrected it. His discoveries, as well as current research on dinosaur feathers and bird development, continue to push the frontier of our understanding.

Opposite

125 **The dinosaur that changed everything in 1969.**

Deinonychus antirrhopus **(cast)**
Cretaceous Period
(115 million years ago)
Carbon County, Montana

The Toothed Birds of Kansas

Early evidence for birds' true origin was uncovered in the 1870s when the Peabody's O.C. Marsh described two ancient birds from Kansas: *Ichthyornis* and *Hesperornis.*

Surprisingly, both had sharp pointed teeth—a very reptilian feature that is rare as hen's teeth today.

Although certain similarities between birds and reptiles had already been noted, the announcement of Marsh's toothed birds attracted the attention of many scholars.

126 **"Restoration of (Ichthyornis) Dispar" Cartoon of O.C. Marsh celebrating with his toothed bird *Ichthyornis dispar,* sent to Marsh by Thomas Nast in thanks for Marsh's gift of a copy of the monograph of his fossils.**

When Marsh first analyzed *Ichthyornis,* he initially thought that the bones were from two different animals: a tern-sized bird and the toothed jaws of a reptile. Marsh quickly recognized this error, realizing that the fossils were from a single animal—an early toothed bird.

Marsh published descriptions of *Ichthyornis* and *Hesperornis* in his 1880 monograph (opposite). Within its pages are bone-by-bone descriptions accompanied by lavish illustrations.

127, 128 **Toothed bird and a close up (inset) of a fossilized lower jaw showing the teeth in place.**

Ichthyornis dispar
Cretaceous Period
(80 million years ago)
Scott County, Kansas
(cast of type specimen)

129 **The classroom in the original Peabody Museum building included a reconstruction of Marsh's toothed bird *Hesperornis* (top center) as well as side-by-side illustrations of a modern bird (a cassowary) and the smallest known dinosaur at the time, *Compsognathus* (far right). Taken together, the similarities of birds and dinosaurs argued for close evolutionary ties.**

Not long after publishing his work on toothed birds, O.C. Marsh received a letter from Charles Darwin—held today within the Peabody's archives. The satisfaction that it provided is evident by the words chosen.

> "I received some time ago your very kind note of July 28th, & yesterday the magnificent volume. I have looked with renewed admiration at the plates, & will soon read the text. Your work on these old birds & on the many fossil animals of N. America has afforded the best support to the theory of evolution, which has appeared within the last 20 years. The general appearance of the copy which you have sent me is worthy of its contents, and I can say nothing stronger than this.
>
> With cordial thanks, believe me yours very sincerely,
>
> Charles Darwin."

130 **Letter from Charles Darwin to O.C. Marsh, dated August 31, 1880, and Marsh's monograph *Odontornithes: A monograph on the extinct toothed birds of North America* (1880).**

Previous page

131 **Fishing the shallow sea that covered Kansas during the Late Cretaceous Period, *Hesperornis* swam through the waters like modern loons or grebes.**

Toothed bird
Hesperornis crassipes
Gove County, Kansas
Cretaceous Period
(80 million years ago)

Montana's "Terrible Claw"

Not long after Marsh's death, Danish artist Gerhard Heilmann argued that the similarities between birds and dinosaurs were superficial, proposing that birds evolved from a more primitive type of reptile.

But in 1969 Peabody paleontologist John H. Ostrom would describe a new dinosaur from the hills of Montana. He named it *Deinonychus,* or "Terrible Claw."

Ostrom's *Deinonychus*—with its avian-like anatomy—reignited conversations and, ultimately, the link between dinosaurs and birds.

132 **A cast of the reconstructed foot of *Deinonychus* showing the positioning of its sickle-like claw bone.**

133 **Restoration of a hypothetical proavian, originally published in *Vor Nuvaerende Viden om Fuglenes Afstamming* (1916) by Gerhard Heilmann (1859–1946).**

134 **John Ostrom plasters a tenontosaur fossil at this 1962 excavation site.**

While studying the anatomy of *Deinonychus* and fossils of the earliest known bird *Archaeopteryx*, Ostrom noted many similarities in their skeletons. Among them was a small but important clue: the "half-moon shaped" wrist bone of *Deinonychus* was very similar to that of *Archaeopteryx*, allowing them both to fold their hands underneath their forearms—just like modern birds.

135 **John Ostrom's photograph of the famed Berlin specimen of the 150-million-year-old early bird *Archaeopteryx* from Bavaria, Germany. O.C. Marsh was offered this specimen for Yale in the spring of 1877.**

Inset

136 **Half-moon shaped wrist bone *Deinonychus antirrhopus* Cretaceous Period (115 million years ago) Carbon County, Montana**

Dinosaurs Deconstructed: Feather Colors and Chicken Beaks

With birds' ancestry now firmly linked to a lineage of small meat-eating dinosaurs, today's Peabody scientists are unraveling the story of this evolutionary transition. Continuing the Peabody's legacy in understanding the evolution of birds, Museum scientists are actively studying the evolution of avian anatomy and plumage.

The feathers of these South American birds show structural (blue) and pigmentary (purple) plumage colors. But do we know the colors of dinosaurs? For some that had feathers, we do!

In 2010 Peabody curators and their colleagues were able to reconstruct the color of the avian dinosaur *Anchiornis*. But how? In birds today, microscopic color-based structures called melanosomes are tied to feather color. In fossilized feathers, preserved melanosomes can be compared to those of modern birds, creating a direct link to color.

_137_ Reconstruction of *Anchiornis huxleyi*, with colors based on research by Peabody scientists.

_138_ Pompadour cotinga
Xipholena punicea
French Guiana, South America

Plum-throated cotinga
Cotinga maynana
Peru, South America

139 **Skulls of an unaltered chicken (left) and alligator (right), with a modified chicken embryo showing a dinosaurian snout (center).**

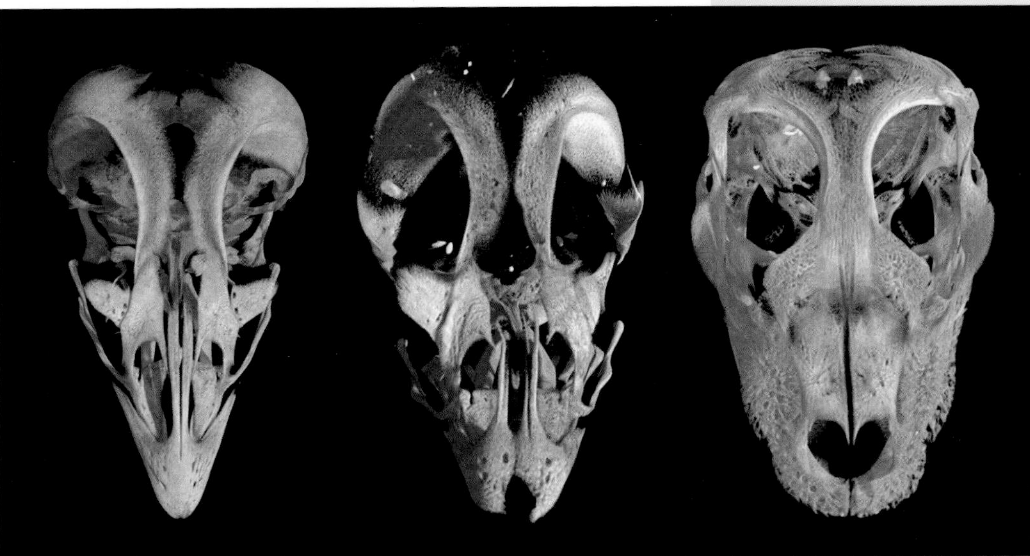

Based on the work of Ostrom and (today) the Peabody's Jacques Gauthier, the evolutionary link between small meat-eating dinosaurs and modern types of birds—like the 25-million-year-old sandpiper shown here—is well received within the scientific community. But other research is now decoding the genetics of this transition.

A 2015 study led by Peabody scientist Bhart-Anjan S. Bhullar brought new insight into the birth of the beak. By blocking the activity of two proteins within a developing embryo, Bhullar and his team produced chicken embryos with more of a reptilian "snout" than bird beak.

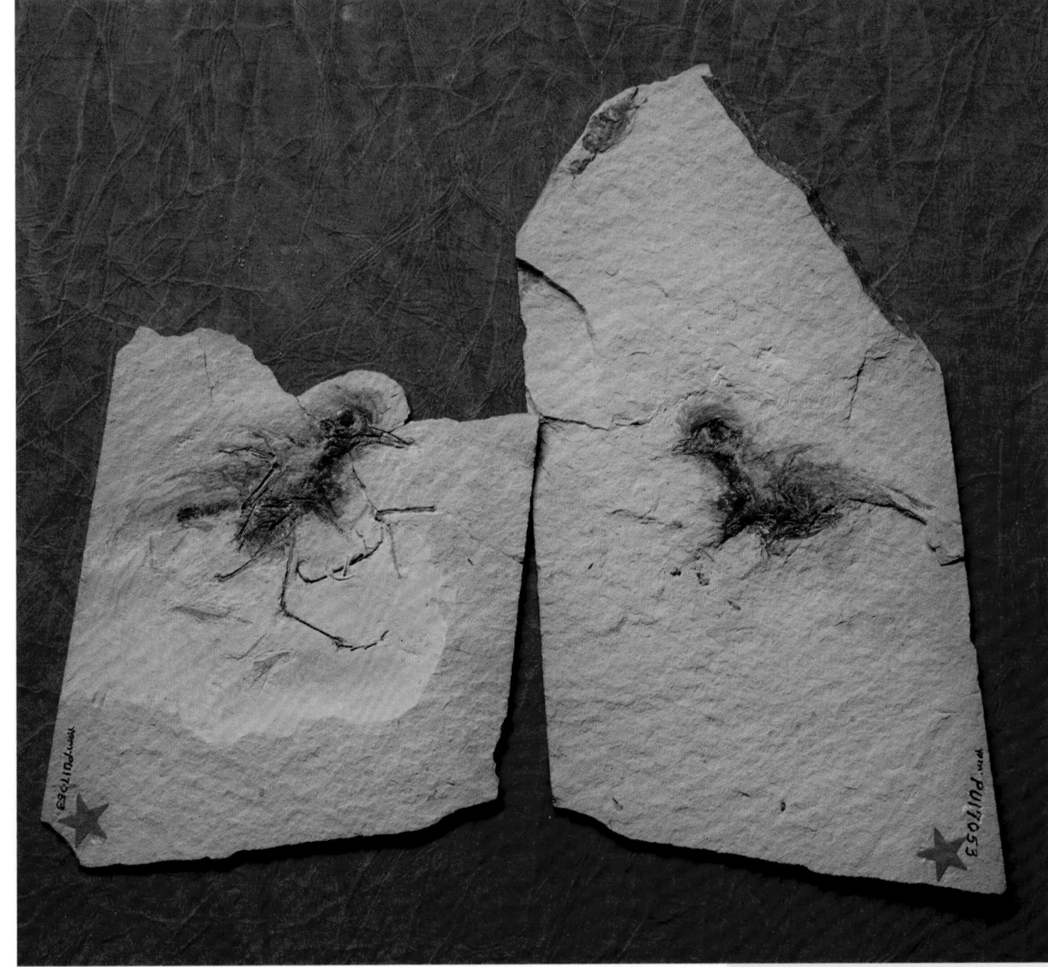

140 **Fossil sandpiper with preserved feathers Oligocene Epoch (25 million years ago) Madison County, Montana**

AQVILA
CAPVT SERPEN
SAGITTARIVS
SCORPIVS
CORONA
ARCTVR9
CAVDA LEONIS
LIBRA
VIRGO
CANCER
LEO
CORVVS
SPICA VIRGINIS
CAPRA
PROCION
CANIS MAIOR
HIDRA
COR LEONIS
PATERA

6

One in a Million

The Peabody Museum is home to more than 13 million objects. Together, these collections document the history of Earth and its peoples. And they help scientists across the globe answer questions about Earth's past, present, and future.

Presented here is an assortment of individual highlights from across the Museum's vast holdings.

Sometimes beautiful, often poignant, these objects and their stories speak to not only the importance of the Peabody's collections, but of those in natural history museums around the world.

Opposite

141 **Astrolabes were used by astronomers and navigators to locate the positions of the planets and stars, and for surveying and triangulation.**

This astrolabe is the oldest object in the Peabody's collection of historical scientific instruments. Constructed by German engineer and astronomer Georg Hartmann, it is one of only four existing brass astrolabes manufactured by Hartmann's workshop in 1537.

Georg Hartmann
astrolabe
1537
Nüremberg, Germany

142 **Inflated caterpillars and the tool used to inflate them.**

143 **Examples of insects with structural color.**

Among the critical aspects of any collection is the preservation of objects. For entomologists, caterpillars provide particular challenges because of their delicate nature and complicated forms.

In practice since at least the late 1800s, the method of inflating dead caterpillars produces immaculate results. Using a simple wheat straw for inflation within a small oven, both the animals' original color and anatomy are preserved exquisitely for generations of scientists.

From camouflage to signaling, animal coloration has been—and continues to be—a popular topic in biology. Although some coloration is due to pigments, other coloration is structural.

The colors of some bird feathers and insects, like those shown here, are structural: microstructures on the surface interfere with light to produce brilliant, often iridescent colors.

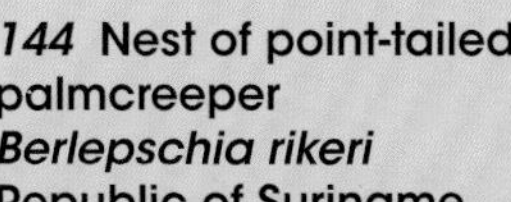

***144* Nest of point-tailed palmcreeper**
Berlepschia rikeri
Republic of Suriname

Collected in South America in 2007, this nest is the first known from the point-tailed palmcreeper—a type of ovenbird. More broadly, it represents the evolutionary transition from cavity nesting to building domed (cavity-like) nests in ovenbirds.

The quagga is an extinct zebra that lived in South Africa until the 19th century. The Peabody's skeleton is that of the only quagga photographed while alive, and only one of seven known quagga skeletons in the world.

The animal was a mare acquired by the Regent's Park Zoo, London in 1851. It died in July 1872 and the Museum's O.C. Marsh purchased the specimen in 1873.

Bone tissue samples from this specimen provided DNA for a 2005 study of zebra relationships. The research showed that the quagga was a geographic variant, or subspecies, of the widely distributed plains zebra.

145 **The Peabody's quagga mare in life in 1870. Only five photographs of this animal are known, taken between 1863 and 1870.**

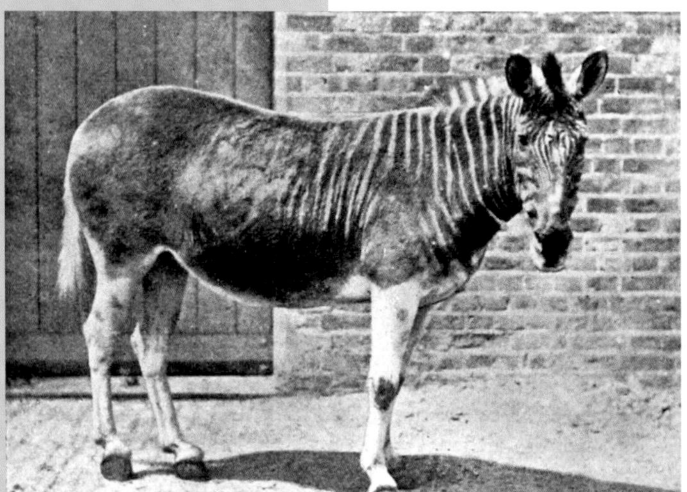

146 **Quagga skull**
Equus quagga quagga
South Africa

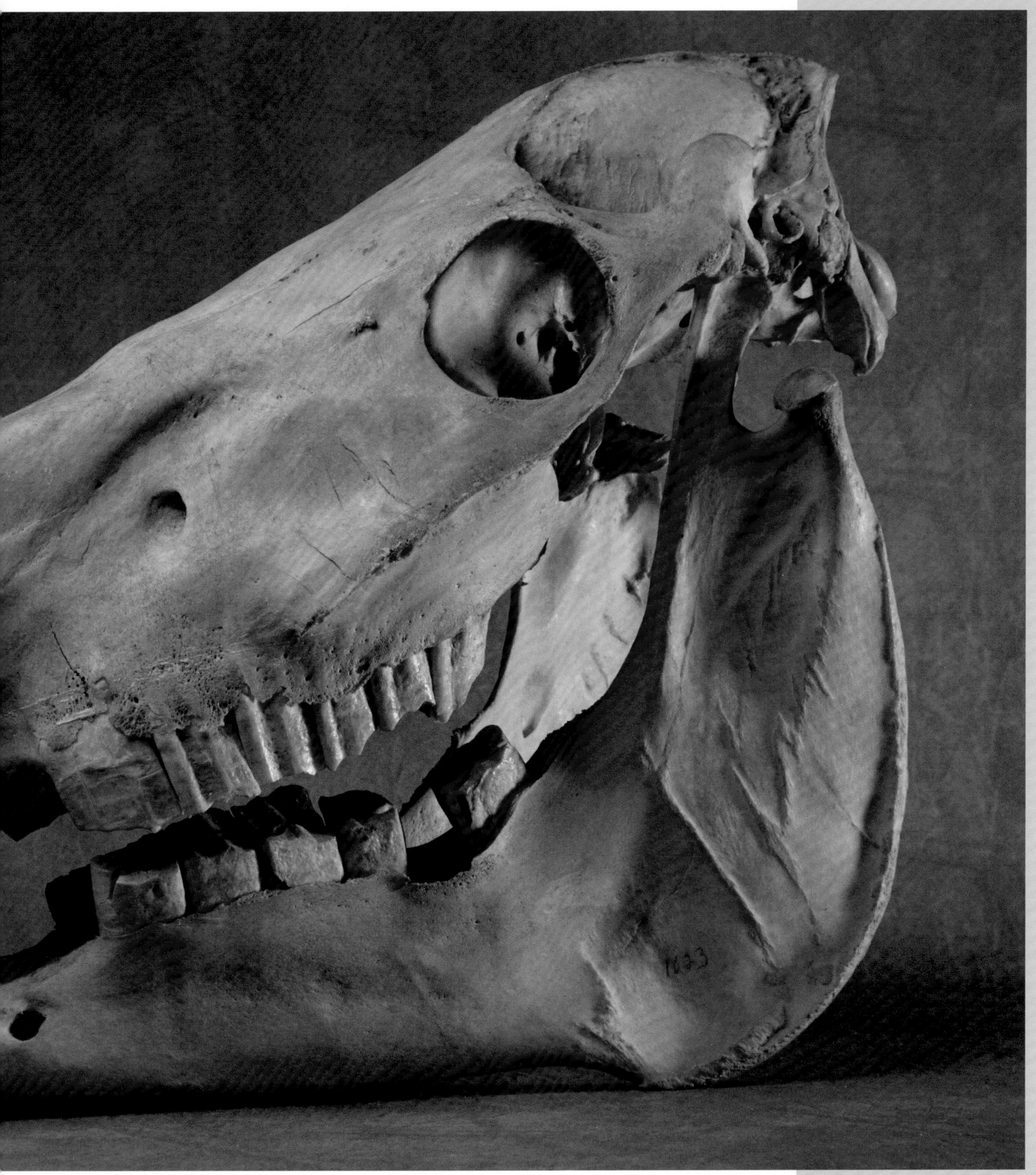

Commonly known as sea lilies, crinoids are close relatives of sea stars and urchins. This slab of *Uintacrinus* highlights the beauty of extraordinary fossil preservation—even the smallest details are preserved.

Still found in marine waters today, crinoids have existed on Earth for more than 450 million years.

The ginkgo and dawn redwood are often described as living fossils: living species very similar to forms found in the fossil record. Native to China today, both plants date back tens of millions of years to the Age of Dinosaurs. Before its rediscovery in the 1940s, the dawn redwood was thought to be extinct!

148 **Fossil dawn redwood branch with leaves**
Metasequoia occidentalis
Oligocene Epoch
(30 million years ago)
Madison County, Montana

Opposite

147 **Fossil sea lily**
Uintacrinus socialis
Cretaceous Period
(85 million years ago)
Logan County, Kansas
(type specimen)

149 **Fossil ginkgo leaf**
Ginkgo adiantoides
Cretaceous Period
(67 million years ago)
Dawson County, Montana

Within the Peabody's holdings is the largest and most diverse collection of eurypterid fossils in the world. Eurypterids, or "sea scorpions," are extinct arthropods closely related to spiders and horseshoe crabs. They thrived for more than 200 million years, until their disappearance during a mass extinction 250 million years ago.

Most eurypterids could be cradled in your arms, but this single tergite (skeletal plate) belonged to a eurypterid the size of a fully grown person!

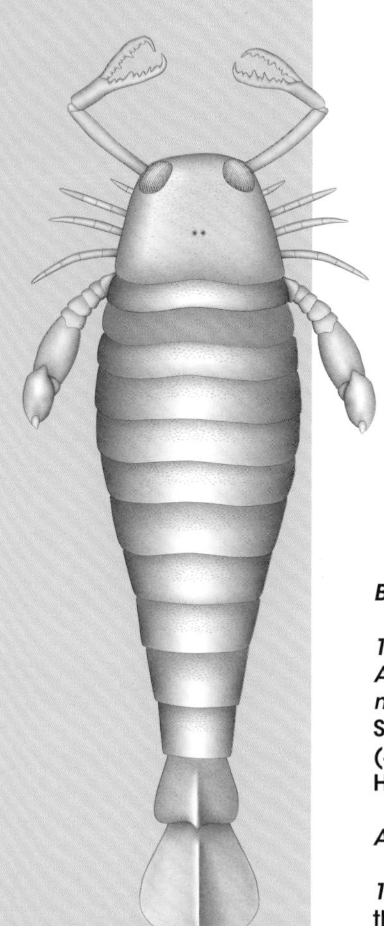

Below

150 **Sea scorpion plate**
Acutiramus macrophthalmus
Silurian Period
(425 million years ago)
Herkimer County, New York

At left

151 **Estimated location of the fossilized plate shown against the full body.**

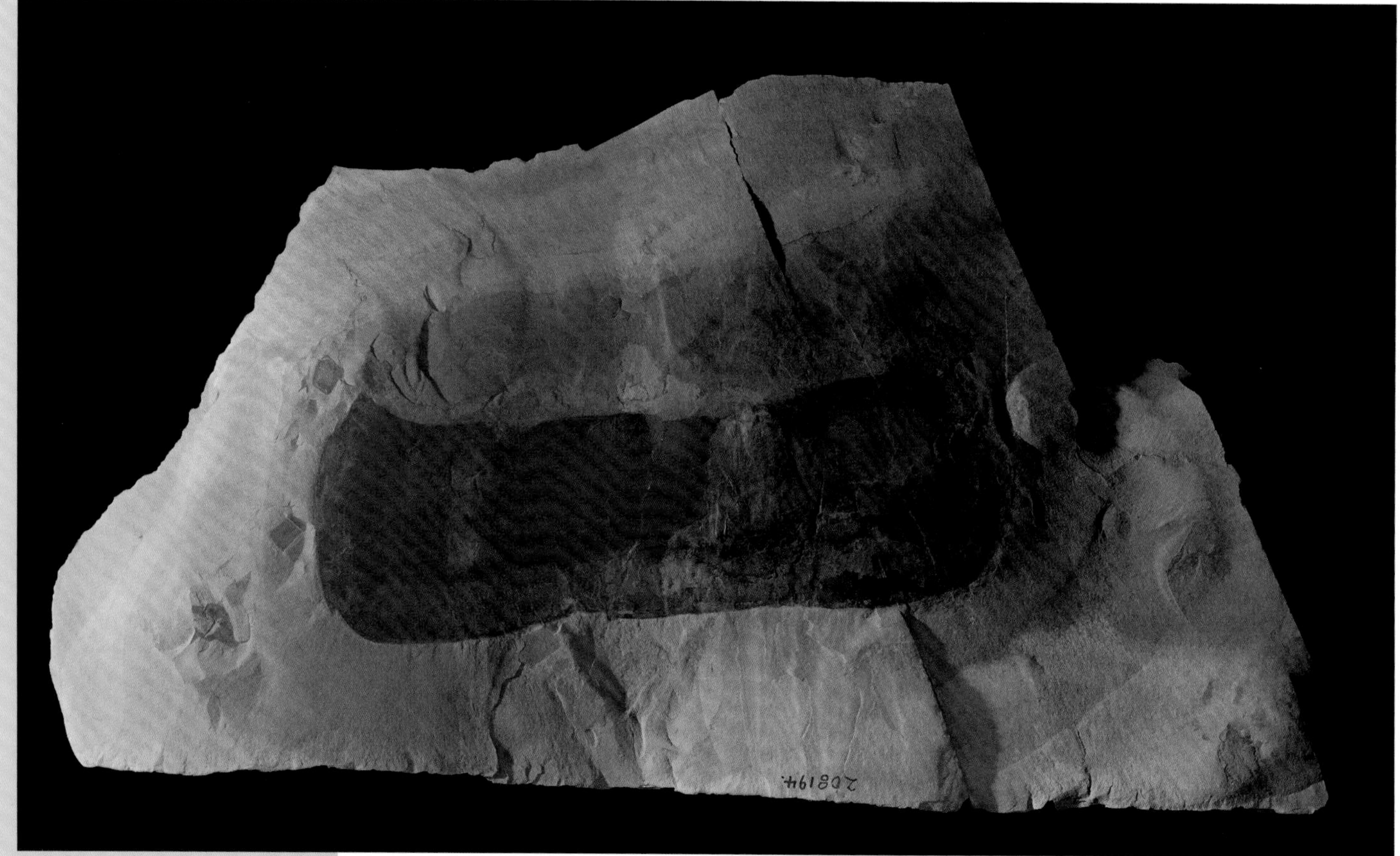

Discovered in 2003, this specimen, thought to be a new kind of primitive crocodile-like reptile, is more than 200 million years old. But unlike crocodiles today, this animal was fully terrestrial and had long limbs like the early dinosaurs that lived alongside it.

Continued study of specimens like this will help us better understand this ancient arms race, and perhaps why dinosaurs would ultimately emerge victorious.

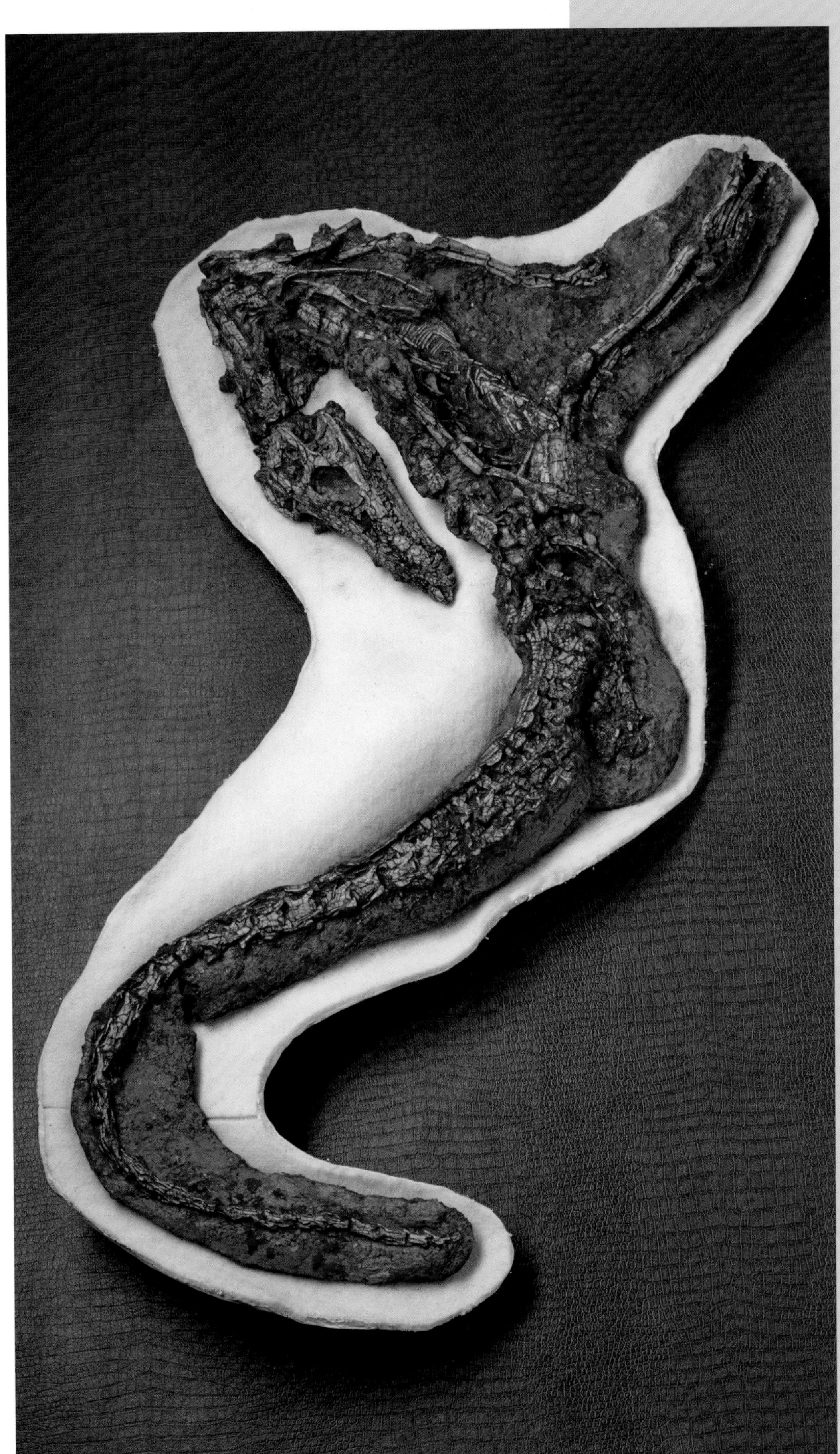

152 **Crocodile-like reptile**
Crocodylomorpha
Triassic Period
(220 million years ago)
Garfield County, Utah

Unique among all animals with backbones, the blackfin icefish lacks both red blood cells and hemoglobin, the protein that carries oxygen throughout the body. Instead it tranports dissolved oxygen within its plasma.

Current work by the Peabody's Thomas Near is helping us understand the evolution of this one-of-a-kind fish.

Relatives of pill bugs, giant isopods like *Bathynomus* live in the deep, cold waters of the Atlantic, Pacific, and Indian oceans. This specimen was collected in the Gulf of Mexico and acquired for the Peabody during an expedition in 2014.

153 **Blackfin icefish**
Chaenocephalus aceratus
Southern Ocean, Antarctica

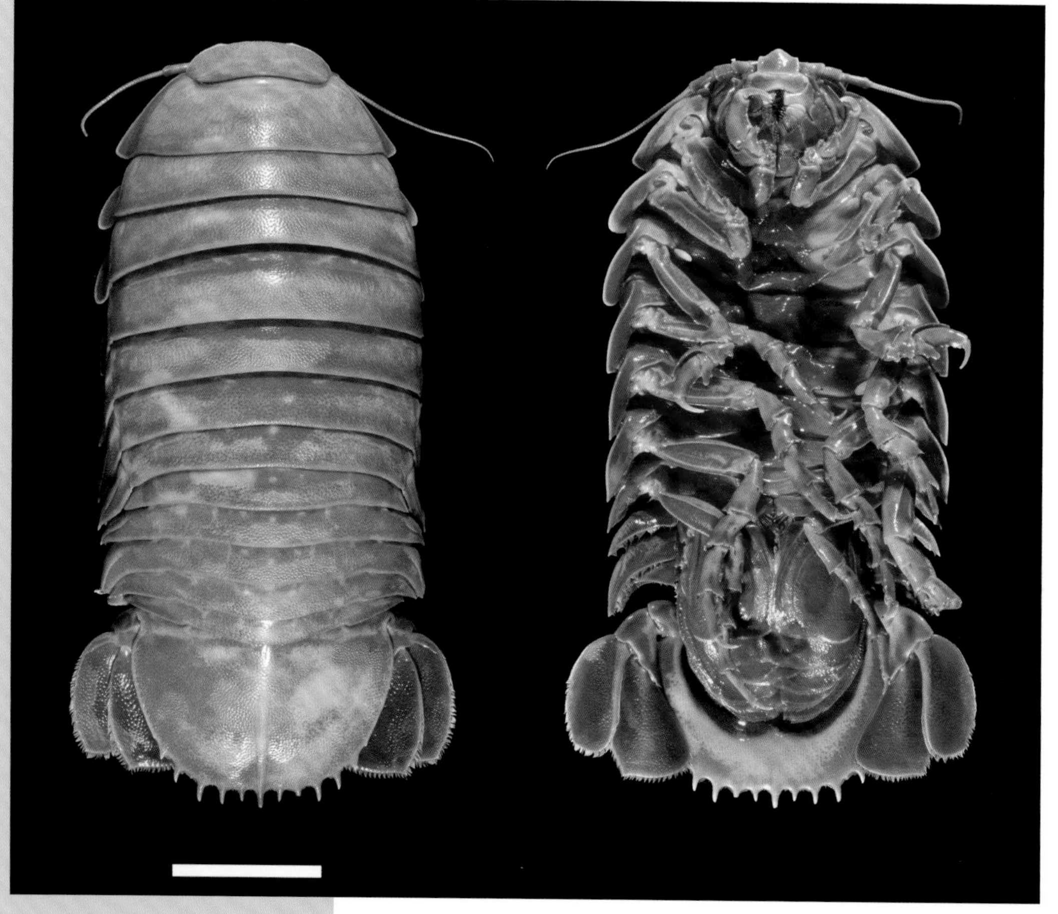

154 **Giant isopod**
Bathynomus giganteus
Gulf of Mexico

The scale bar equals 5 centimeters (about 2 inches).

Known from warm shallow waters of the South Pacific and Indian oceans, *Tridacna gigas* is the largest living bivalve. Capable of living for more than 100 years, its shells can reach four feet (more than a meter) and together weigh more than 500 pounds (225 kilograms)!

Today giant clams are nearly endangered due to overfishing for food and shells.

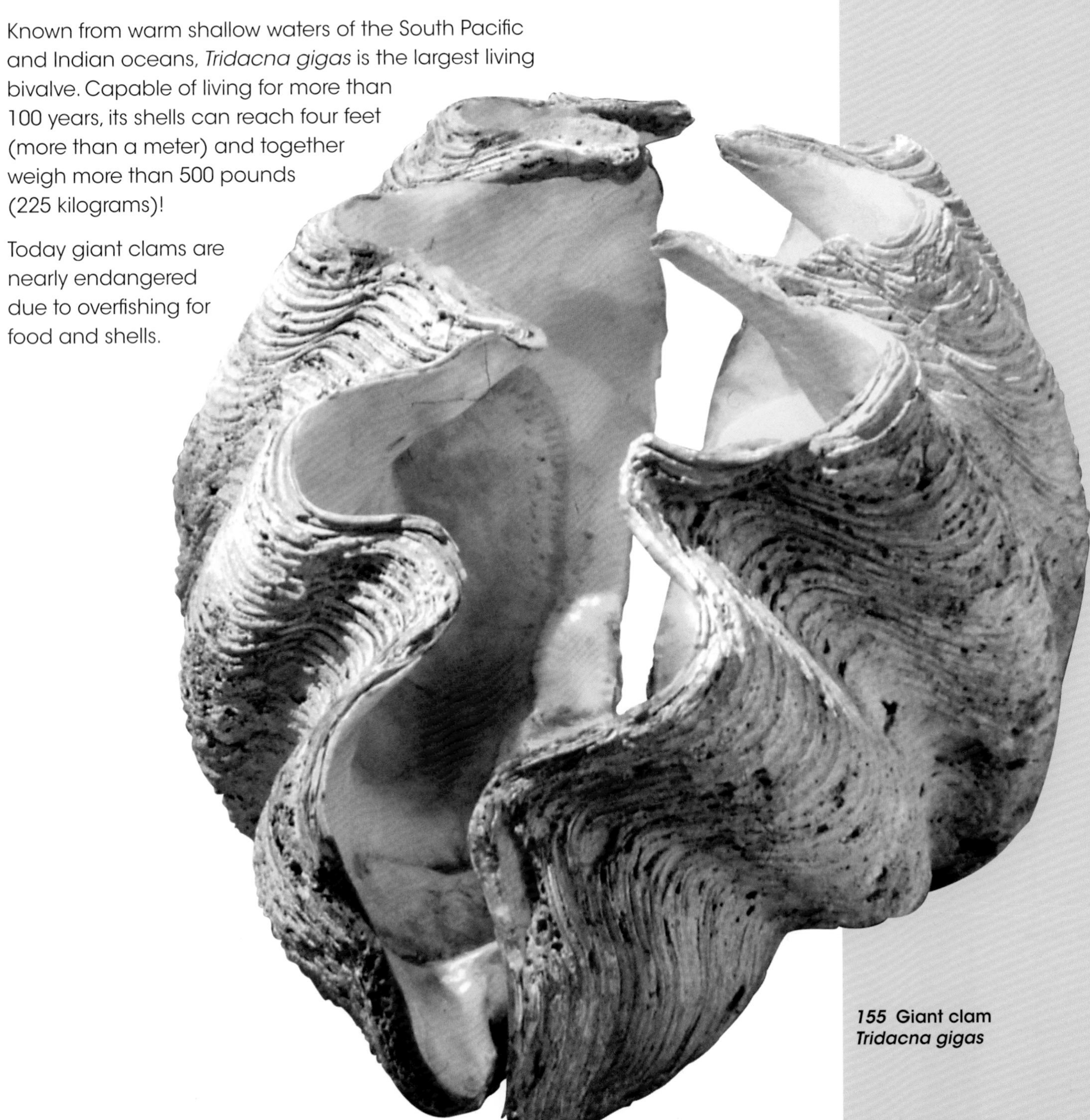

155 **Giant clam**
Tridacna gigas

CARRIER
LUMBER

156 Double-cylinder electrical machine ca. 1770

The first electrostatic generators date to the mid-17th century. Over time they were modified to increase efficiency. With its double glass cylinders, this machine dates to the 18th century and was possibly used by Yale's seventh president Ezra Stiles.

Opposite

157 Vacuum chamber from 27-inch (69-centimeter) cyclotron made by Ernest O. Lawrence (1901–1958), who earned his doctorate at Yale and received the 1939 Nobel Prize in Physics for his invention.

158 **Charles Darwin**

This liverwort specimen was collected by Charles Darwin during the five-year voyage of the HMS *Beagle*. Liverworts are small, non-vascular plants that are found across the globe.

During the expedition Darwin collected many specimens and recorded many observations, which would ultimately lead to the publication of his *Origin of Species* in 1859.

Although the timing of its collection is not certain, Darwin also collected this sea fan and had it sent to the Peabody's James Dwight Dana.

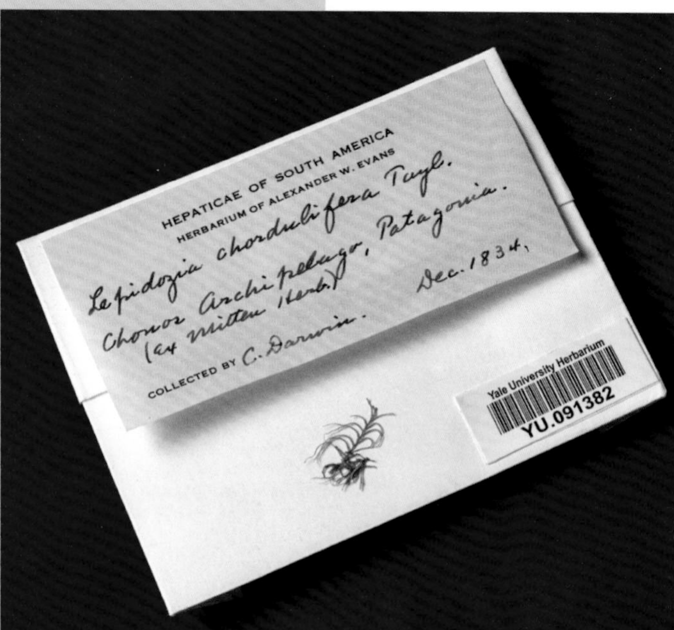

159 **Leafy liverwort collected by Charles Darwin in 1834**
Lepidozia chordulifera
Chonos Archipelago, Chile

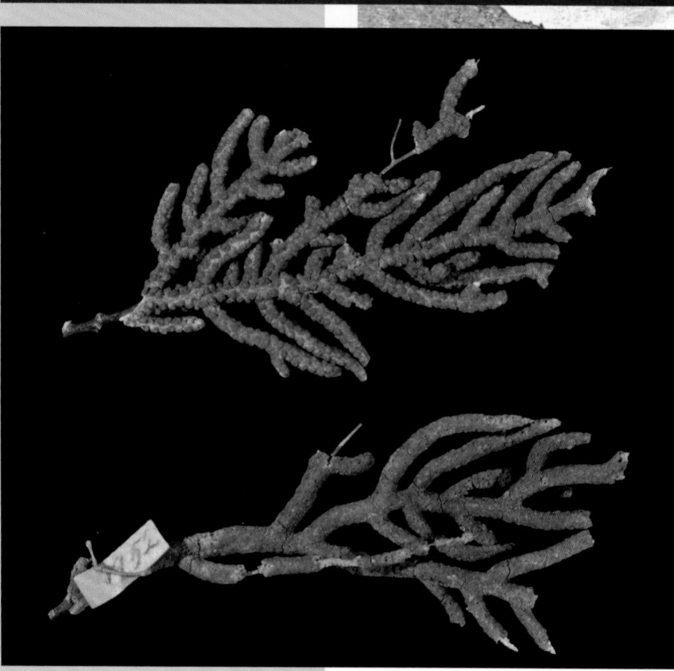

160 **Sea fan**
Lophogorgia sanguinolenta
Locality not specified

161 **"HMS *Beagle* in the Straits of Magellan"**
Frontispiece from *Journal of researches into the natural history and geology of the various countries visited by H.M.S. Beagle etc.* by Charles Darwin
First Murray illustrated edition, 1890
Artist, Robert Taylor Pritchett (1828–1907)

In 1876 Thomas H. Huxley, the famous English biologist known as "Darwin's Bulldog" for his defense of evolution, visited New Haven to meet O.C. Marsh. Impressed with Marsh and the Peabody, he wrote, "there is no collection of fossil vertebrates in existence which can be compared with it."

In this cartoon—drawn for Marsh during his visit—Huxley predicted how an ancestral horse might appear and showed an imaginary extinct human riding it. Huxley named his creations "*Eohomo*" (dawn human) and "*Eohippus*" (dawn horse). Marsh would name one of his fossil horses *Eohippus* later that year.

162 **Thomas H. Huxley in 1874, two years before his visit to New Haven.**

Line engraving
Artist, C.H. Jeens

163 **T.H. Huxley's fanciful cartoon and George R. Wieland's copy of O.C. Marsh's monograph on fossil horses.**

"*Eohippus* & *Eohomo*"
(1876)
Artist, Thomas H. Huxley
(1825–1895)

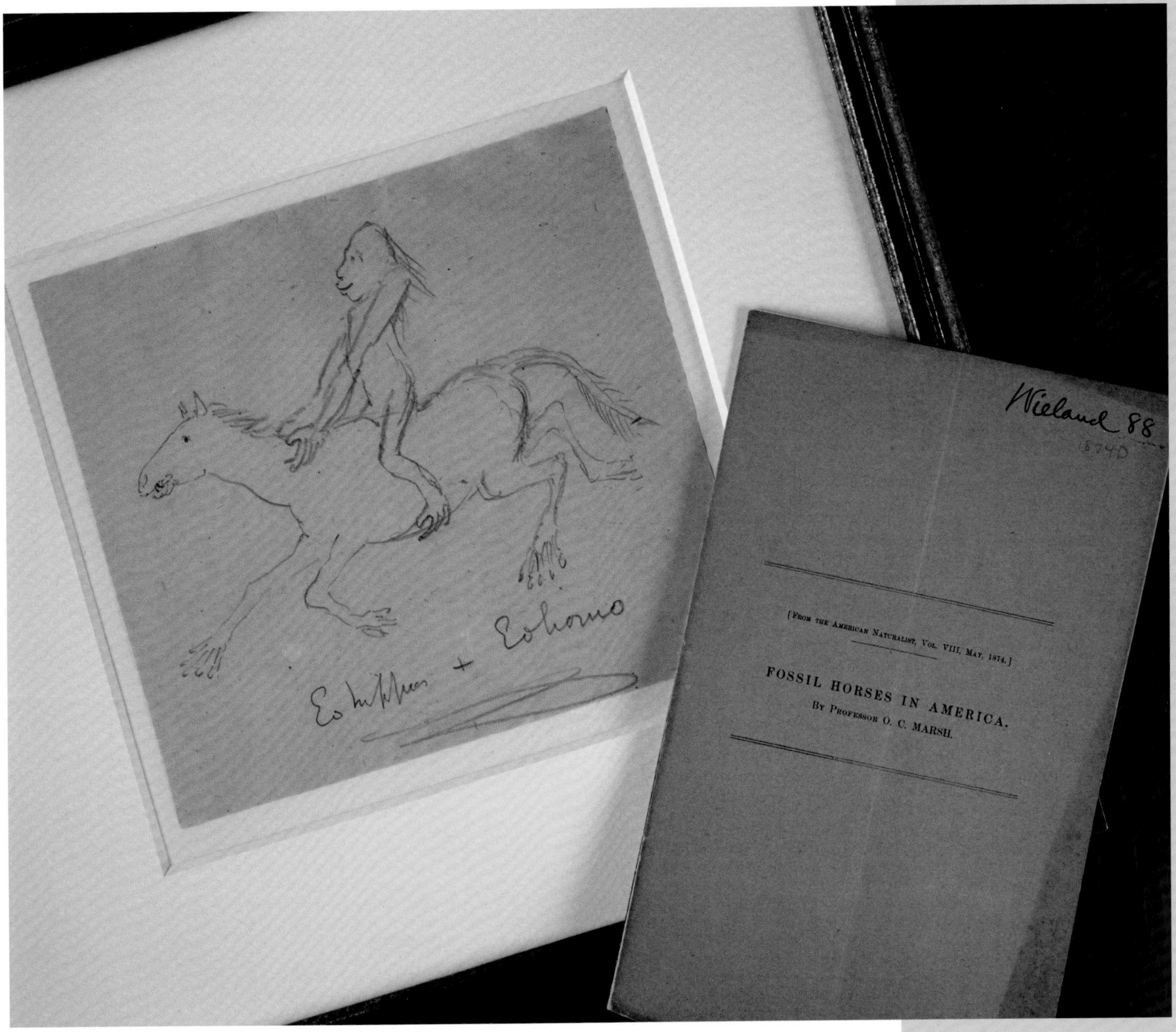

Carved from a single piece of wood and standing 46 inches (17 cm) tall, this debating stool depicts a standing male figure with inlaid shell eyes. Debating stools are used by the Iatmul people of the middle Sepik River region of Papua New Guinea.

Never used for sitting, the stools are used during formal debates. The debates occur in the presence of the village's most prominent ancestor, depicted on the stool's back.

165 **Michael Coe and one of the kneeling ballplayer sculptures excavated from the site of San Lorenzo.**

The Olmec civilization of southern Mexico originated around 1500 BCE and thrived for more than 1,000 years. Most commonly associated with the Olmec are colossal heads sculpted from basalt.

Shown here is an exact, fiberglass copy of the San Lorenzo Colossal Head 6, made in 1967 by a team of archaeologists led by the Peabody's Michael Coe. Standing at five and one-half feet (almost two meters), Head 6 is one of the smaller sculptures—though it weighs up to 10 tons—and likely depicts an Olmec king.

In the 1960s, Coe and colleagues used magnetometry—mapping patterns of magnetism in soils—to locate many of the colossal heads.

166 **Olmec head (cast)**
1200–900 BCE
San Lorenzo, Veracruz

Opposite

164 **Debating stool**
Mid-20th century
Papua New Guinea

FERNS OF NORTH AMERICA. 293

PLATE XXXIX.—FIG. 1–6.

NOTHOLÆNA SINUATA, KAULFUSS.

Wavy-leaved Notholæna.

NOTHOLÆNA SINUATA:—Root-stock short and thick, very ... with narrow ferruginous scales; stalks short, covered, ... young at least, with ciliated scales; fronds six inches to ... feet high, rigid, narrowly oblong-linear, simply pinnate; ... numerous, short-stalked, coriaceous, roundish or ovate, ... somewhat cordate, obtuse, nearly entire or sinuated or ... lobed; upper surface more or less sprinkled with stel... or pinnately divided white scales; lower surface and rachis ... covered with ferruginous ovate or ovate-lanceolate cili... scales which conceal the sub-marginal sporangia.

... sinuata, KAULFUSS, Enum. Fil., p. 135.—KUNZE, in Linnæa, xiii., p. 135; Die Farrnkraüter, p. 95, t. 45.—LINK, Fil. Hort. Berol., p. 145.—MARTENS & GALEOTTI, Syn. Fil. Mex., p. 46. —LIEBMANN, Mex. Bregn., p. 61.—FÉE, 8me Mém., p. 117; ...me Mém., p. 12.—METTENIUS, Fil. Hort. Lips., p. 45.—EATON, in Bot. Mex. Boundary, p. 234.—HOOKER, Sp. Fil., v., p. 107 (excluding var. bipinnata); "Bot. Mag., t. 4699."—HOOKER & BAKER, Syn. Fil., p. 370.—FOURNIER, Pl. Mex., Crypt., p. 120. ...dium sinuatum, SWARTZ, Syn. Fil., p. 14.—WILLDENOW, Sp. Pl., v., p. 120. ...gramme sinuata, PRESL, Tent. Pterid., p. 219.—METTENIUS, Cheil-

167 Amos Eaton's first herbarium, four volumes (1815–1816)
Author, Amos Eaton (1776–1842)

168 Daniel Cady Eaton (1834–1895) was the grandson of Amos Eaton and the first curator of the Yale Herbarium. This plate from D.C. Eaton's beautifully illustrated book *Ferns of North America* depicts (left to right) *Notholaena sinuata, N. newberryi* and *N. ferruginea.*

Despite an interest in natural history, Amos Eaton pursued a career in the Catskills area of New York as a land agent and surveyor. Accused of forgery during a land dispute, Eaton was imprisoned for five years.

While in jail Eaton devoted his time to studying the natural sciences. On his release in 1815, he studied at Yale under Benjamin Silliman and others.

Eaton soon began his herbarium, including plants from New York and New England. Composed of four volumes containing around 90 sheets each, this historic herbarium is today housed in the Peabody's botanical collections.

169 **Porcelain rice-beer jar**
Ming Dynasty, China (1300–1600 CE)
Province of Ifugao, Philippines

With its blue-on-cream glaze, this rice-beer jar was collected by the Peabody's Hal Conklin during his studies of rice agriculture in the Philippines. Rice-beer jars of this kind were highly valued heirloom pieces in remote agricultural communities of the Philippine highlands.

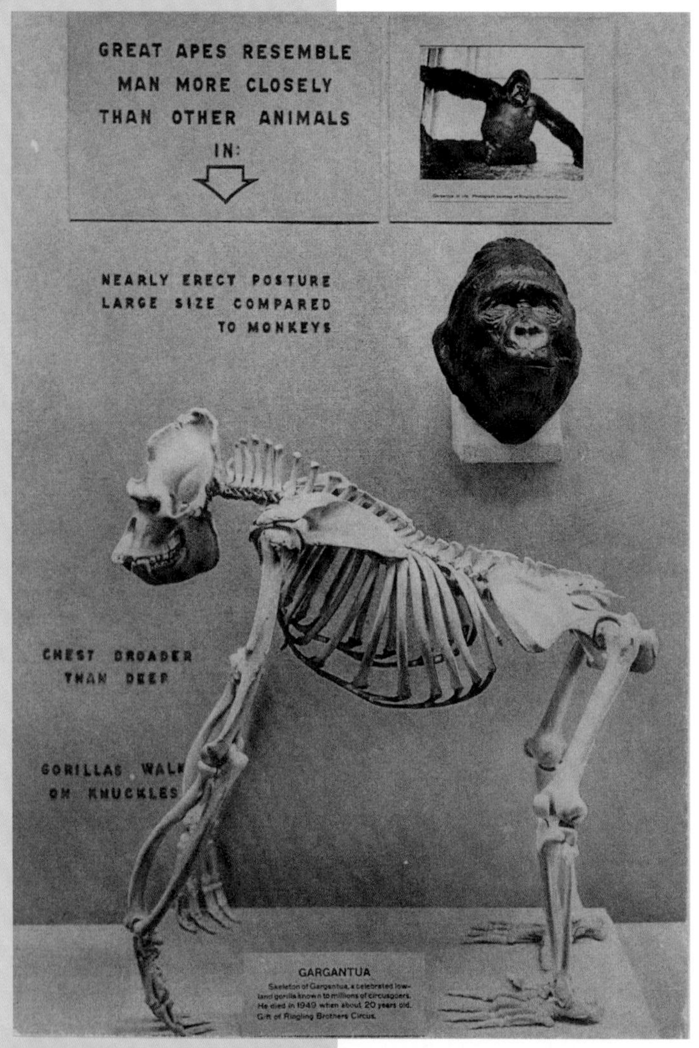

The famed gorilla Gargantua the Great was orphaned when he was only a month old and lived with missionaries in Africa. In 1931 a sea captain purchased him and brought him to Boston.

In 1937 he was offered to the Ringling Bros. and Barnum & Bailey Circus. A delighted Henry Ringling North (Yale '33) named him Gargantua the Great. Millions stood in line to see the lowland gorilla.

Gargantua died of double pneumonia in November 1949 at the age of 20. North donated Gargantua's skeleton to the Peabody in 1950.

170 **This postcard from 1951 shows the skeleton of the famous lowland gorilla from Ringling Bros. and Barnum & Bailey Circus as originally displayed, with a model of a gorilla head and a photograph of Gargantua in life.**

Gargantua the Great
Gorilla gorilla
Carbon County, Montana

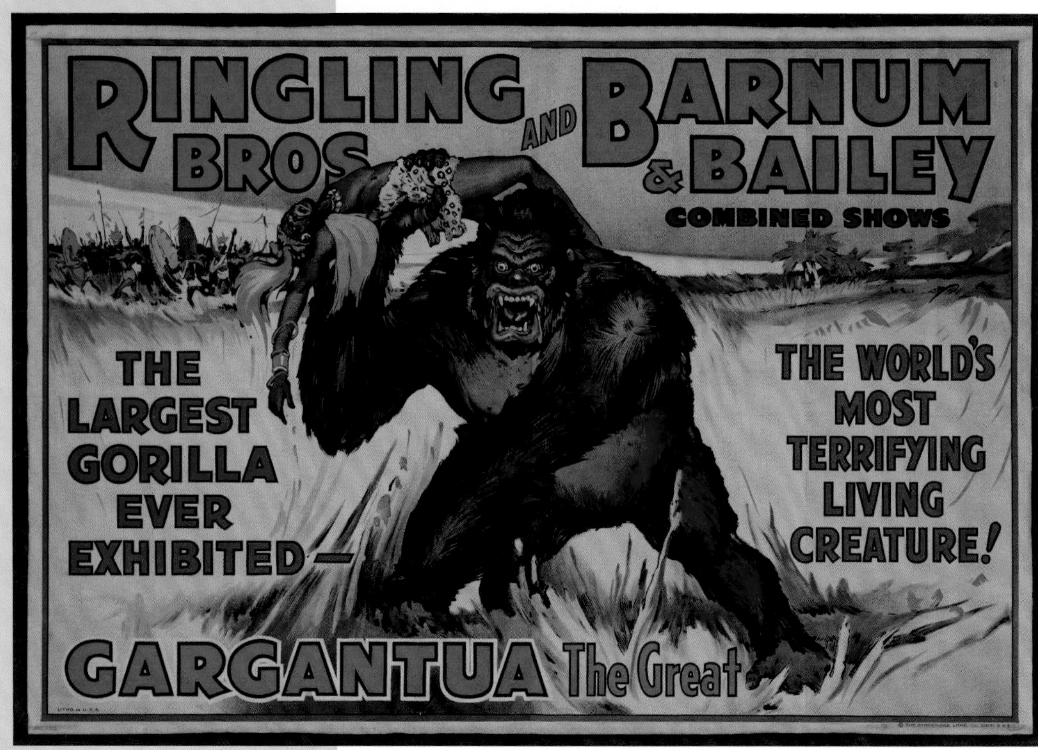

171 **Supported by flamboyant promotion, Gargantua toured the nation for 12 years.**

"Ringling Bros. and Barnum & Bailey: Gargantua the Great" circus poster
Ink on paper
No date
The Strobridge Lithographing Company

Once one of the largest woodpeckers in the world, the ivory-billed woodpecker is presumed extinct. Though sightings have been reported, no concrete evidence exists to support their survival today.

These specimens show the differences in bill shape between the young and adults. The discovery of this delayed bill maturation—first noted in Peabody specimens—has important implications for the ecology of this species: young birds had limited foraging abilities and depended on their parents for several months after leaving the nest.

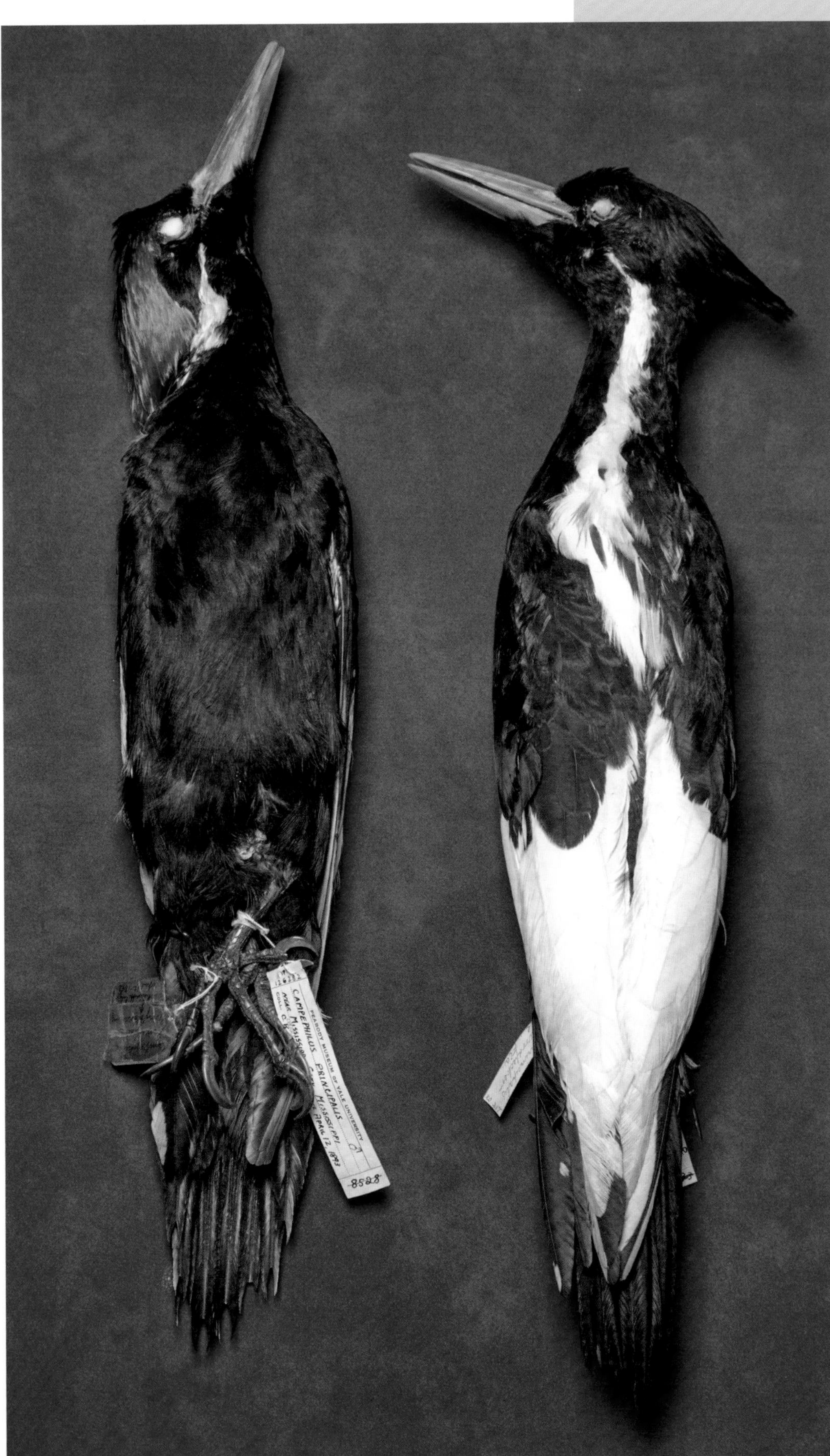

***172* Ivory-billed woodpecker, immature female (right)**
Campephilus principalis principalis
Osceola County, Florida

Ivory-billed woodpecker, adult male (left)
Campephilus principalis principalis
Florida

Recovered in 1772, the Krasnojarsk meteorite is the first pallasite meteorite discovered. Pallasites are a rare type of stony-iron meteorite—only 105 are known.

Krasnojarsk was the first meteorite to reveal a meteorite texture now known as the Widmanstätten pattern—long interwoven crystals that form during cooling.

173 **Krasnojarsk meteorite Krasnoyarsk Territory, Russia**

Inset

174 **An example of Widmanstätten patterns found within a meteorite.**

Perhaps the most celebrated of all dinosaurs, *Tyrannosaurus rex* was named three years after a fairly complete skeleton was discovered in Montana in 1902. But this specimen—a single tooth—is the first fossil ever collected of the tyrant king.

Discovered in 1874 by a student on a field trip, it was sent to the Museum, where—because of the sheer volume of fossils being collected at the time—it was left unstudied for decades. The tooth was finally identified as belonging to *T. rex* in 2002.

175 Tyrannosaurus rex
tooth
Cretaceous Period
(67 million years ago)
Jefferson County, Colorado
(shown twice actual size)

7

A Changing Earth

Our Earth is a dynamic planet. Continents move, climates change, and species evolve and go extinct.

Natural history collections document this complex history. And they are also records of the modern world, allowing us to see change happening today.

From the study of fossils to ancient cultures, we are able to better understand the changes that we all face today.

Opposite

176 **Fossil palm frond with fishes**

Sabalites **sp.**
with *Priscacara* sp. (large)
and *Knightia* sp.
Early Eocene
(50 million years ago)
Lincoln County, Wyoming

Change in the Past

Throughout its 4.6-billion-year history, Earth has continually evolved. As the great plates shift, continents move and seas open or close.

Over time the composition of the atmosphere has also fluctuated, with the amount of oxygen, carbon dioxide, and other gases changing over time. These changes affect global climate, producing cool periods, such as the ice ages, and warm ones, with no permanent ice at either pole.

All throughout the Age of Dinosaurs, it was a much warmer world. This warmth continued after the extinction of the dinosaurs, peaking around 50 million years ago when global temperatures averaged around 82 degrees Fahrenheit (28 °C), compared with 59 °F (15 °C) today.

Warmer temperatures shift life's distribution, allowing animals adapted to warmer climates to move farther toward the poles. Fifty million years ago palms, alligators, and other warmth-adapted species lived as far north as New Haven.

***177* Fossil alligator**
Alligator* cf. *A. prenasalis
Oligocene Epoch
(30 million years ago)
Custer County,
South Dakota

Atmospheric concentration of the greenhouse gas carbon dioxide is of particular interest to scientists studying climate change today.

In 2014 Peabody curator Jay Ague and collections manager Stefan Nicolescu showed for the first time that enormous amounts of carbon dioxide can be released from Earth's crust at subduction zones, where two of Earth's plates converge and one is thrust beneath the other.

Chemical analysis of this specimen from Greece was key to their study.

178 **Metacarbonate rock**
Marlas, Greece

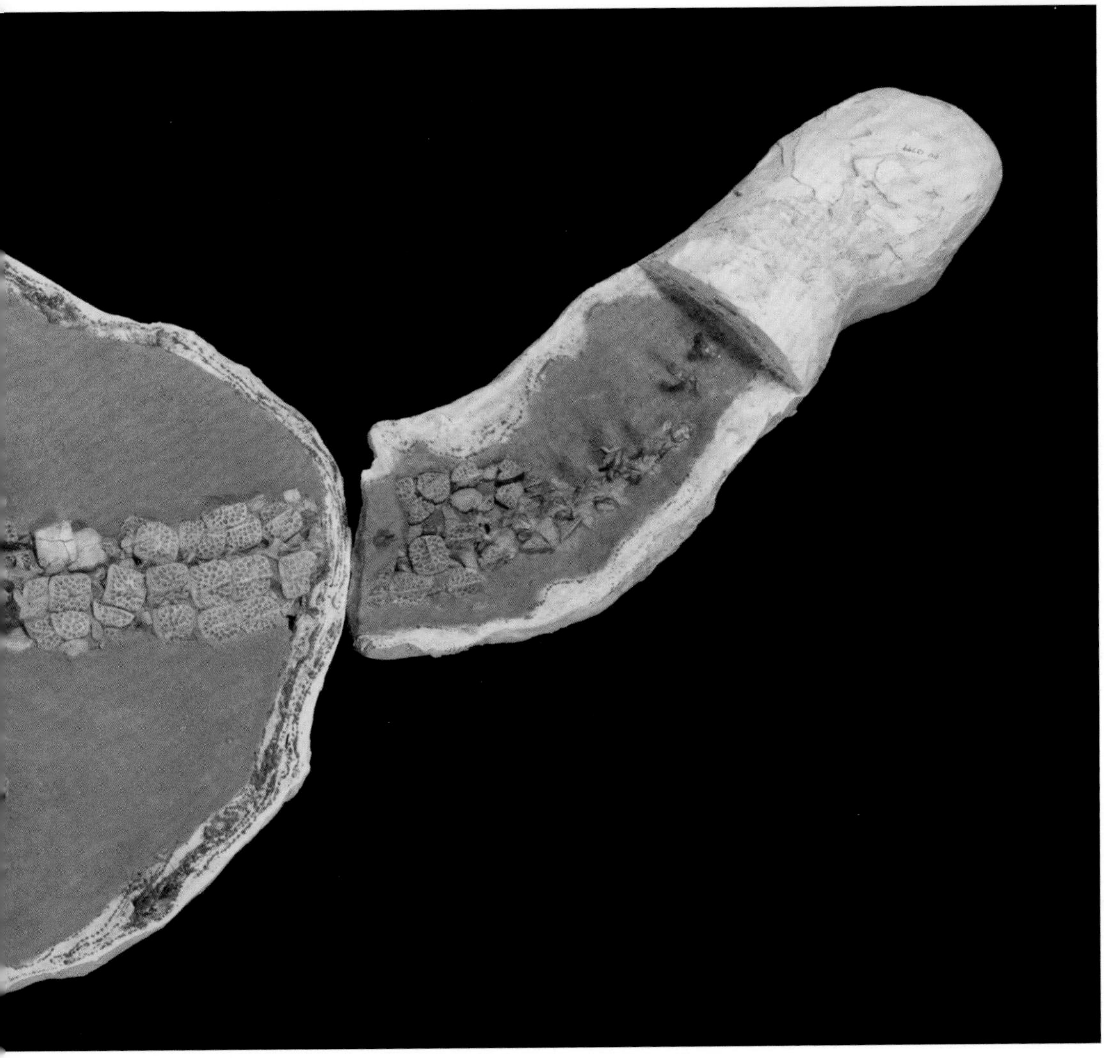

179 Three-pointed stone (zemi) with carved face
ca. 1200–1500 CE
Dominican Republic

180 Shell pendant
2000 BCE–500 CE
Haiti

_181_ Saladoid bowl
ca. 250 BCE–500 CE
Virgin Islands

Changing Populations

With more than 130,000 objects, the Peabody's Caribbean archaeological collection is the largest in the world. Most of these artifacts were collected by (Ben) Irving Rouse (1913–2006), who studied in the Caribbean for nearly 70 years.

Rouse was deeply interested in the distribution of culture over space and time. While Rouse's archaeological research focused on the Caribbean, where he established the chronology of human occupation of the islands and adjacent South America, his methods for documenting movement of people and technology can be applied to other places, as well.

Rouse's chronological framework for the human occupation of the Caribbean is still in use today.

Human Population and Change Today

The human population reached seven billion in 2011, growing more than five billion in the last century. United Nations projections predict eight billion as soon as 2020.

As humanity continues to expand, so do its requirements for habitat and resources.

As historical documents, natural history collections are providing scientists with critical information for understanding the consequences of human actions and growth.

They also show us that these changes are occurring at a rate far faster than in Earth's past—within a lifetime—and within our own backyards.

182 **Look closely. This frog has not two but three hind legs. Research by Peabody scientists has linked limb deformities with the development of landscapes by humans. While parasites may lead to some deformities, the specific causes for most remain unknown.**

Green frog
Rana clamitans melanota
New Haven County, Connecticut

183 The area of Sachem's Wood as it was in the 19th century.

Detail from "The City of New Haven, Conn. 1879" Artists, O.H. Bailey and J.C. Hazen ca. 1879.

184, 185 The elliptic-leaved shinleaf occurs in forested habitats. These specimens were collected by James D. Dana in 1851 in Sachem's Wood, right next to the Peabody's current home. Today Sachem's Wood is characterized by manicured lawns, scattered trees, buildings, and sidewalks. As a result of this habitat destruction, the plant (inset) thrives there no more.

Elliptic-leaved shinleaf
Pyrola elliptica
New Haven County, Connecticut

186 **Asian shore crab**
Hemigrapsus sanguineus
Fairfield County, Connecticut

Scale bar equals 25 millimeters (about 1 inch).

The Asian shore crab (*Hemigrapsus sanguineus*) is native to East Asia. But from North Carolina to Maine, reports of this invasive species became more common in the 1990s. Collected by Peabody scientists in 1998, specimens that are the first recovered from Connecticut are now housed within the Peabody's collection. Today, Asian shore crabs are abundant along the shores of Connecticut. The specimen shown here was collected as part of a workshop for local teachers in 2013. In less than an hour, nearly 150 specimens were recovered.

As carbon dioxide emissions rise and global temperatures increase, animals and plants respond. By comparing observations today with historical data from specimens within the Peabody's collections, we also see that leafing out times of many deciduous trees in New England are, as average spring temperatures rise, now occurring earlier than they were 100 years ago.

A Brilliant Future

The stories of natural history are endless. They are the stories of humanity and its world, and they are worth celebrating.

For the past 150 years, the collections and researchers of the Yale Peabody Museum have contributed greatly to the advancement of science. And for decades and centuries to come they will continue to be among the most significant in shaping our understanding of the natural world, and of our place within it.

Opposite

187 **Black oak**
Quercus velutina
New Haven County, Connecticut

HERBAR
OF
YALE UNIVERS

PLYMOVENT
CONSERVATION

Acknowledgments

This book was developed in conjunction with the Yale Peabody Museum's 150th anniversary and the exhibition *Treasures of the Peabody: 150 Years of Exploration & Discovery.* To celebrate such a rich history requires the knowledge and efforts of many talented individuals. It has been a pleasure to work with such enthusiastic and capable colleagues and friends. We thank the exhibition design team of Laura Friedman, Sally Pallatto, and Kimberley Zolvik, as well as Rosemary Volpe for photography art direction and with volunteer Corrie Roe for the layout of the catalog. We owe a particular debt to Robert Lorenz, whose fantastic photographs bring the Peabody's story to life. Richard Conniff, at work on his own wonderful book on the Peabody, has been both an inspiration and a great sounding board. Walter Brenckle, Robert Charlesworth, John Ferro, Angeliki Manthou, and Nicole Palffy-Muhoray constructed the exhibition, while Michael Anderson and Maishe Dickman produced artifact and specimen mounts. David Heiser offered pedagogical guidance; we are thankful to Harry Shyket for his technical and online support, Rich Boardman for security, and Catherine (Cap) Sease for her conservation work. We are also grateful to Eliza Cleveland, Linda Warner, and Sung Yun for their fundraising efforts, and to Melanie Brigockas for the development of marketing materials. For their logistical support, we thank Bonnie Mahmood and Sharon Rodriguez, as well as Susan Castaldi, Angie Chirico, Lynn Jones, and Nate Utrup. The Museum's Director of Collections and Operations Tim White played a variety of roles, including content development, coordinating object loans, and overseeing much of the catalog photography. We offer special thanks to Richard Kissel, Director of Public Programs, for his leadership and coordination of the exhibition's development and production, as well as his editorial work on the catalog.

The exhibition (and by extension this book) would have been impossible without the invaluable contributions of the curators and staff across the Museum's ten scientific divisions, who provided insight into their collections and the histories of the divisions: Anthropology (Rod McIntosh, Richard Burger, Oswaldo Chinchilla, John Darnell, Michael Dove, Anne Underhill, Roger Colten, Becky DeAngelo, Erin Gredell, and Maureen DaRos White); Botany (Michael Donoghue and Patrick Sweeney); Entomology (Leonard Munstermann, Ray Pupedis, and Larry Gall); Historical Scientific Instruments (Paola Bertucci); Mineralogy and Meteoritics (Jay Ague, Stefan Nicolescu, and Barbara Narendra); Paleobotany (Peter Crane and Shusheng Hu); Invertebrate Paleontology (Derek Briggs, Susan Butts, and Jessica Utrup); Vertebrate Paleontology (Jacques Gauthier, Eric Sargis, Bhart-Anjan Bhullar, Chris Norris, Dan Brinkman, and Marilyn Fox); Invertebrate Zoology (Leo Buss, Eric Lazo-Wasem, Dan Drew, and Lourdes Rojas); and Vertebrate Zoology (Rick Prum, Jacques Gauthier, Eric Sargis, Greg Watkins-Colwell, and Kristof Zyskowski). Thanks to Bill Landis, Christine McCarthy, and the Manuscripts and Archives staff at Sterling Memorial Library, and Amy Dowe and Mark Mitchell at the Yale University Art Gallery. Finally, we offer our deepest thanks to Peabody archivist Barbara Narendra, whose mentorship on Peabody history has been invaluable to this project.

— David Skelly and Tom Near

Senior Conservator Catherine Sease reinforces tears on the back of a tapa, or barkcloth, with Japanese tissue and wheat starch paste in the Peabody's Conservation Laboratory. The Conservation Laboratory actively supports and promotes the Museum's mission to preserve and protect the collections entrusted to its care. The museum conservator is involved in all aspects of collections care and handling, from collections conservation to storage to exhibition installation. Because the Peabody collections are used primarily for research, treatment is conservative and the care of specimens is guided by the principle that the integrity of an object should be preserved in every possible way.

Tapa cloth
Inner bark of mulberry tree
Mid 20th century
Samoa, Polynesia
Brandt collection

21.2.5.3e
YRM 1981

Credits

Numbered entries indicate objects and other materials that are shown in this catalog. Photography by Robert Lorenz unless otherwise noted.

Front cover (left to right)

Othniel Charles Marsh (1831–1899)
Oil on canvas
1982
Artist, Rudolph Franz Zallinger (1919–1995), American
YPM YPMAR 000740
Donor, Rudolph F. Zallinger, 1982

Pyromorphite
Phoenixville, Chester County, Pennsylvania
Digital photograph
2004
Photographer, William K. Sacco/Yale Peabody Museum
YPM MIN 055574
Collector, unknown

Replica of Colossal Head 6, Olmec (*see 166*)

Bear Seamount coral with brittle star (*see 102*)

Culpeper/Loft double microscope, "The Yale Microscope" (*see 1*)

17-year periodic cicada
Magicicada septendecim
1877
New Haven, New Haven County, Connecticut
YPM ENT 451314
Collector, Oscar Harger, May 24, 1877

page ii

House poles and figures from the Malsin Collection acquired in the 1970s from the Sepik River region in New Guinea are stored in this Anthropology collections storage room, which houses a portion of the Peabody's extensive Oceania Collection. Systematic collections from Oceania include ethnographic materials collected by Yale faculty and graduate students conducting research in New Guinea, the Philippines, Malaysia, and the Solomon Islands.

Totem figures
Oceania Collection
Division of Anthropology
Kline Geology Laboratory
Yale University

Chapter 1 Exploration, Discovery, and Understanding

1 Culpeper/Loft double microscope, "The Yale Microscope"
Wood, brass, shark skin, cardboard
1735
YPM HSI 020001
Purchased by Yale College, 1735

2 "New Haven Palisaded or Fortified"
From *A Pictorial History of Raynham and Its Vicinity*
1900
Author, Charles Hervey Townshend
Courtesy of the New Haven Museum

3 Portrait of Gov. Elihu Yale (1649–1721)
Oil on canvas
1717
Artist, Enoch Seeman the younger (1694–1744), British
Yale University Art Gallery, Gift of Dudley Long North, M.P., 1789.1

4 "A View of the Buildings of Yale College at New Haven"
Left to right: Union Hall (South College), First Chapel (Athenaeum), South Middle College (now Connecticut Hall), Connecticut Lyceum, and Berkeley Hall (North Middle College). Only Connecticut Hall still stands. Students playing football on the New Haven Green.
Lithograph
1807
Artist, Amos B. Doolittle (1754–1832)
Yale University buildings and grounds photographs, 1716–2004 (inclusive). Manuscripts & Archives, Yale University; RU 0703, drawer 21, folder 70, image no. 2081

5 Miniature of Benjamin Silliman
Oil on ivory
ca. 1815
Artist, Nathaniel Rogers (1788–1844), American
Yale University Art Gallery, Gift of Miss Maria Trumbull Dana, 1954.34.1

6 Hematite
Cumbria, England
YPM MIN 052719
Acquired by Benjamin Silliman, 1805

7 Kelp ashes
ca. 1817
YPM HSI 290001
Sent to Benjamin Silliman from Normandy by Gay Lussac

8 *American Journal of Science and Arts*
First Series, Volume 1, Issue 1
YPM YPMAR 000877

9 Weston meteorite
Stony, ordinary chondrite
Fairfield County, Connecticut
December 14, 1807 (recorded fall)
YPM MIN 010305
Gibbs Cabinet, purchased 1825

"An Account of the Meteor"
Memoirs of the Connecticut Academy of Arts and Sciences
1810
Authors, Benjamin Silliman and James L. Kingsley
YPM YPMAR 000897

10 Beryl, variety emerald
Colombia
YPM MIN 053131
Gigot d'Orcy collection, 18th century

Topaz
Saxony, Germany
YPM MIN 054401
Gigot d'Orcy collection, 18th century

Illustrated plates from *Histoire Naturelle*
Beryl, variety emerald, plate LVII
Topaz, plate LIV
ca. 1783–1789
Facsimiles
Author, Jean Fabien Gautier d'Agoty (1747–1781), F.L. Swebach Desfontaines

11 James Dwight Dana (1813–1895), B.A. 1833, M.A. 1836
Oil on canvas
1858
Artist, Daniel Huntington (1816–1906), American
Yale University Art Gallery, Bequest of Edward Salisbury Dana, B.A. 1870, 1961.46

12 *A System of Mineralogy,* first edition
New Haven: Durrie & Peck and Herrick & Noyes
1837
Author, James Dwight Dana (1813–1895)
YPM MINAR 000112
Author's personal copy, signed

Rock hammer
Metal, wood
"JDD" on handle
YPM MINAR 000132
Owned by James Dwight Dana

Crystal model of calcite
Glass
ca. 1836
YPM MINAR 000135
Made by James Dwight Dana

13 Othniel Charles Marsh (1831–1899)
Photographic print
ca. 1856
Photographer, unknown
YPM VPAR 000038
Gift of the Estate of Othniel Charles Marsh, 1899

14 George Peabody (1795–1869)
Oil on canvas
Date, artist unknown
Photographer, unknown
Gift of the Estate of Othniel Charles Marsh, 1899 (YPM YPMAR 000898)

15 "Mr. Peabody's Letter" and "The Instrument of Gift"
Original Documents of the Peabody Museum of Natural History, Yale University
October 22, 1866
YPM YPMAR 000878

16 Members of the Yale College Scientific Expedition of 1870
Standing, left to right: John Wool Griswold, Henry Bradford Sargent ('71 S., M.A. '07), George Bird Grinnell ('70, Ph.D. '80), Charles Wyllys Betts ('67, M.A. '71), Othniel Charles Marsh, Charles T. Ballard, John Reed Nicholson ('70), James Matson Russell ('70).

Only a fraction of the massive collection of dinosaur fossils begun under O. C. Marsh in the 19th century are on public display. The rest is stored beneath the Museum. These fossils include the first described specimens of many famous dinosaur species, such as *Apatosaurus* and *Stegosaurus*. With support from the National Science Foundation and a Save America's Treasures grant through the Institute of Museum and Library Services, skilled Vertebrate Paleontology Preparation Lab staff and volunteers spent hundreds of hours repairing and conserving hundreds of bones, making it possible for this scientific treasure to be available for research to scientists and students from around the world today and into the future.

O.C. Marsh Collection
Division of Vertebrate Paleontology
Kline Geology Laboratory
Yale University

Foreground, left to right: Eli Whitney ('69, M.A. '72), Alexander Hamilton Ewing ('69), Harry Degen Ziegler ('71), Bill the cook (far right)
Photographic print
1870
Near Fort Bridger, Wyoming
Photographer, unknown
YPM YPMAR 002775

17 Addison Emery Verrill (1839–1926)
Oil on canvas
1910
Artist, John Henry Niemeyer (1839–1932), American, b. Germany
YPM IZAR 005917
Photographer, unknown

18 Beef tapeworm
Taenia saginata
Removed from student, March 18, 1896
New Haven, New Haven County, Connecticut
YPM IZ 023806
Acquired by Addison E. Verrill, 1896

Round worm
Ascaris lumbricoides
Picked from vomit at railroad depot
New Haven, New Haven County, Connecticut
YPM IZ 038073
Collector, J. Gilbin, December 7, 1879

19 George Jarvis Brush (1831–1912), Ph.B. 1852, M.A. (Hon.) 1857
Oil on canvas
Date unknown
Artist, Harry Ives Thompson (1840–1906), American
Yale University Art Gallery, 1912.9

20 Proposal for the first Peabody Museum
Architectural rendering
Watercolor and graphite on paper
ca. 1873
Artist, J. Cleaveland Cady, architect
YPM VPAR 002473

21 Hugh Gibb with *Brontosaurus*, first Peabody Museum building
Photographic print
ca. 1902
Photographer, unknown
YPM VPAR 000096

22 The first Peabody Museum
Photographic print
Undated
Photographer, unknown
YPM VPAR 000186

23 Display with Otisville mastodon, first Peabody Museum building
Mammut americanum (YPM VP 012600)
Photographic print
Undated
Photographer, unknown
YPM VPAR 000165

24 George R. Wieland with *Archelon ischyros*, first Peabody Museum building
YPM VP 003000 (type specimen)
Photographic print (digitally restored by William K. Sacco)
November 3, 1914
Photographer, F. H. Simonds
YPM YPMAR 002776

25 "Professor Marsh's Primeval Troupe. He Shows His Perfect Mastery Over The Ceratopsidœ."
Punch, Or The London Charivari, Vol. 99
September 13, 1890
Artist, possibly Edward Tennyson Reed (1860–1933)
YPM VPAR 002243

26 Mounted *Claosaurus*, first Peabody Museum building
Now *Edmontosaurus* (YPM VP 002182)
Photographic print
ca. 1901
Photographer, unknown
YPM YPMAR 000174

27 Vase
Ceramic
Greco-Roman, Ptolemaic period, 300–100 BC
Elephantine Island, Egypt
YPM ANT 006951
Collector, K. Arnold (Old Collection)

Peabody Museum Ethnology and Archaeology Catalogue
1867
YPM ANTAR 035888

28 Puma effigy mortar
Stone
Cuzco, Peru
Photographer, William K. Sacco/Yale Peabody Museum
YPM ANT 017451
Purchased by Hiram Bingham III, 1922

29 Hiram Bingham III, Yale Peruvian Expedition of 1912
Photographic print
September 1912
Photographer, Ellwood C. Erdis
Yale University/National Geographic Creative

30 Inca ruins at Machu Picchu
Hand-colored slide
1911
Photographer, Harry Ward Foote
Yale Peruvian Expedition papers, 1908–1948 (inclusive). Manuscripts and Archives, Yale University, image no. 3120

31 Peabody Museum proposed addition facing Whitney Avenue
Architectural drawing, May 27, 1930
Artist, Charles Z. Klauder (1872–1938), architect
Yale University buildings and grounds photographs, 1716–2004 (inclusive). Manuscripts and Archives, Yale University, RU 0703, box 40, folder 837, image no. 2294

32 Peabody Museum construction
Photographic print
ca. 1924
Photographer, unknown
YPM YPMAR 000887

33 Yale Tennis Courts
Postcard
ca. 1910
Publisher, Danzinger & Berman, New Haven
YPM YPMAR 000884

34 Dedication exercises, Peabody Museum
Photographic print
December 29, 1925
Photographer, George Keeley, *New Haven Register*
YPM YPMAR 000889

35 Richard Swann Lull with Jurassic diorama
Photographic print
Undated
Photographer, unknown
YPM VPAR 000280
Richard Swann Lull Archives

36 Letter to Clarence Darrow
July 10, 1925
YPM VPAR 000280
Richard Swann Lull Archives

37 Telegram from Clarence Darrow to Richard Swann Lull
July 10, 1925
YPM VPAR 000280
Richard Swann Lull Archives

38 *Brontosaurus* skeleton, Great Hall
Photographic print
1930
Photographer, unknown
YPM YPMAR 000544

39 Richard Swann Lull and Hugh Gibb, Great Hall
Photographic print
1930
Photographer, unknown
YPM YPMAR 000888

40 Untitled
Oil on canvas
1927
Artist, Wilfrid Swancourt Bronson (1894–1985), American
Photographer, Jerry Domian/Yale Peabody Museum
YPM YPMAR 000038

41 Mural fragment (dragonfly)
Detail from Carbonaceous Forest diorama, Invertebrate Hall, Peabody Museum
Fresco
1926
Artist, Robert Bruce Horsfall (1869–1948), American
YPM YPMAR 000891

42 Timber Line Diorama
ca. 1958-1959
Artists, J. Perry Wilson (1889–1976) (background); Ralph C. Morrill (1902–1996) and David H. Parsons (1928–1985) (foreground and animals)
Photographer, Jerry Domian/Yale Peabody Museum
YPM YPMAR 000892

43 Model for the Timber Line diorama
Wood, plaster, latex rubber
ca. 1958
Artist, J. Perry Wilson (1889–1976)
Photographer, Jerry Domian/Yale Peabody Museum
YPM YPMAR 000893

44 *The Age of Mammals*, preliminary drawing
Graphite on paper
1953
Artist, Rudolph Franz Zallinger (1919–1995), American
YPM YPMAR 000880

45 Dry pigments, sable brushes, casein, and "palette"
Artist, Rudolph Franz Zallinger (1919–1995)
Photographer, Jerry Domian/Yale Peabody Museum
YPM YPMAR 000881

46 *The Age of Reptiles*, preliminary drawing
Graphite on paper
1942
Artist, Rudolph Franz Zallinger (1919–1995), America
YPM YPMAR 000879

47 *Deinonychus* quarry site, near Bridger, Montana
Color 35mm slide
ca. 1965
Photographer, unknown

48 Sickle claw bone
Deinonychus antirrhopus
Carbon County, Montana
Cretaceous Period
2000

Photographer, Jerry Domian/Yale Peabody Museum
YPM VP 005205
Collector, John H. Ostrom and Grant E. Meyer, Cloverly Expedition, 1964

49 Reconstruction of *Deinonychus antirrhopus*
Frontispiece from "Osteology of *Deinonychus antirrhopus*, an Unusual Theropod from the Lower Cretaceous of Montana," *Bulletin of the Peabody Museum of Natural History*, Yale University, Volume 30
1969
Artist, Robert T. Bakker (1945–)
YPM VPAR 002530

50 Yale students, laboratory class, Archbold Biological Station, Lake Wales Ridge, Venus, Florida
Digital photograph
October 2012
Photographer, Raymond J. Pupedis/Yale Peabody Museum

51 Wadi el-Hôl, Theban Desert Road Survey, Yale Egyptological Institute in Egypt
Digital photograph
2003
Photographer, John C. Darnell/Yale University

52 Curator Jay Ague, Hammonasset State Park
Digital photograph
2009
Photographer, James Sirch/Yale Peabody Museum

53 The DSV *Alvin* on R/V *Atlantis*, Manning Seamount, New England Seamounts, North Atlantic
Digital photograph
May 30, 2003
Photographer, Jon A. Moore
Used with permission

54 Gwen Antell and Sara Kahanamoku-Snelling, Barbados
Digital photograph
2015
Photographer, Philipp S. Arndt
Used with permission

55 Burrowing Owl, Suriname
Digital photograph
2007
Photographer, Kristof Zyskowski/Yale Peabody Museum

56 Marilyn Fox, skull of a fossil crocodile relative (YPM VP 057103)
Digital photograph
2016
Photographer, Christina Lutz
Used with permission

57 Yale field class, Old Man McMullen Pond, Yale Camp at Great Mountain Forest, Litchfield Hills, Connecticut
Digital photograph
September 2015
Photographer, Raymond J. Pupedis/Yale Peabody Museum

58 Melina Delgado, Division of Paleobotany
Digital photograph
2015
Photographer, Shusheng Hu/Yale Peabody Museum

59 Danica Meier and Max Lambert, preparing specimens (YPM HERR 019418, *Malayopython reticulatus*, on its back, and YPM HERR 016419, *Morelia spilota*)
Digital photograph
2015
Photographer, Gregory J. Watkins-Colwell/Yale Peabody Museum

60 Paleobotany field trip, Block Island, Rhode Island
Digital photograph
2014
Photographer, Cliff Vanover
Used with permission

61 Collecting *Viburnum*, Chiapas, Mexico
Digital photograph
2015
Photographer, Patrick Sweeney/Yale Peabody Museum

62 24-hour Peabody BioBlitz, Stratford, Connecticut
Digital photograph
October 9, 2010
Photographer, Lynn Jones/Yale Peabody Museum

63 Holger Petermann, Petrified Forest National Park, Arizona
Digital photograph
2014
Photographer, Alan Zdinak
Used with permission

Chapter 2
Discovering Nature

64 Lesula monkey
Cercopithecus lomamiensis (type specimen)
Skull, skin
Democratic Republic of Congo
YPM MAM 014080
Collector, Maurice Emetshu, August 12, 2008

65 "Lashley type" saliometer (drool collector)
Owned by Ivan P. Pavlov
1916
YPM HSI 150002.A
Donor, Roberta Yerkes Blanchard, 1988

66 X-ray of a human head
Radiograph
April 26, 1896
Photographer, Arthur W. Goodspeed
YPM HSI 051441A
Donor, Stephen Irons on behalf of the Yale Department of Physics, 2005

67 *Fagopsis longifolia*
(type specimen)
Florissant, Teller County, Colorado
Eocene Epoch
YPM PB 030121
Collector, W.P. Cockerell, June 1907

68 Leopard frog
Rana kauffeldi (type specimen)
Bloomfield, Richmond County, New York
Digital photograph
2011
Photographer, Gregory J. Watkins-Colwell/Yale Peabody Museum
YPM HERA 013217
Collector, Brian R. Curry, November 15, 2011

69 Fairfieldite
Branchville, Fairfield County, Connecticut
YPM MIN 033118
Collector, Abijah N. Fillow, ca. 1876

70 *Apatosaurus ajax*
(type specimen)
Vertebra
Morrison Quarry 10, Jefferson County, Colorado
Jurassic Period
YPM VP 001860
Collector, Arthur Lakes, 1877

71 *Heliobatis radians*
Lincoln County, Wyoming
Eocene Epoch
YPM VP 007261
Collector, unknown

72 *Docodon striatus*
(type specimen)
Como Bluff Quarry 9, Albany County, Wyoming
Jurassic Period
YPM VP 011823
Collector, Othniel C. Marsh, 1880

73 Life-size model of *Architeuthis dux*, 1883
Papier-mâché, length 12.2 m (40 ft)
Designed by Addison E. Verrill (1839–1926)
Artist, James H. Emerton (1847–1931)
Photographic print
Undated
Photographer, unknown
YPM YPMAR 000894

74 *Architeuthis dux*
Beak (left)
Newfoundland, Canada
YPM IZ.010272.GP
Collector, George Simms, December 1874

Tentacle (right)
Logy Bay, Newfoundland, Canada
YPM IZ 009634.GP
Collector, Moses Harvey, November 25, 1873

75 *Opabinia regalis*
British Columbia, Canada
Cambrian Period (Burgess Shale)
YPM IP 005809
Collector, Charles D. Walcott, pre-1921

76 Life-size model of *Architeuthis dux*, 1960s
Plastic foam, steel and fiberglass, length 11.4 m (37.5 ft)
Artists, Henry Townshend, Edward Migdalski, George Rennie, Ralph C. Morrill, and Rollin Bauer
Digital photograph
2004
Photographer, Jerry Domian/Yale Peabody Museum

77 *Aegirocassis benmoulai* in environment
Based on specimen YPM IP 237172
Digital illustration
2015
Artist, Marianne Collins
© 2015 Marianne Collins

78 *Aegirocassis benmoulai*
Zagora–Ezegzaou hill, Morocco
Ordovician Period
YPM IP 237172
Collector, Mohammed Ou Said Ben Moula, 2011

79 Vendobionta assemblage
Cast
Newfoundland
Precambrian
YPM IP 037199
Collector, Adolf Seilacher, August 26, 1990

80 George R. Wieland
Great Hall, Peabody Museum
Photographic print
November 8, 1931
Photographer, unknown
YPM YPMAR 000890

81 Mold (dorsal aspect) of *Triarthrus eatoni*
(YPM IP 000228)
Copper-colored metal
YPM IP 202147
Made by Charles E. Beecher, January 1897

Scrapbook of original drawings of fossil and recent invertebrates
1882
Author, Charles E. Beecher

YPM IPAR 000814
Donated by Nancy Herzig

Double X-ray of *Triarthrus eatoni*
(YPM IP 000228)
Photographic print
1970s
Photographer, John Cisne
YPM IPAR 000557

Triarthrus eatoni
Oneida County, New York
Ordovician Period
YPM IP 000228
Collectors, Charles E. Beecher and F. L. Nason, July 1893

82 Ordovician Beecher Trilobite Bed locality
Cleveland Glen, Rome, Oneida County, New York
Photographic print
ca. 1900
Photographer, unknown
YPM IPAR 000815
Donated by Nancy Herzig

83 "Exhibition of Marine Life at the Seychelles College"
Le Seychellois, Republic of Seychelles
1958
YPM IZAR 005908

84 Willard D. Hartman (1921–2013) with the sponge *Willardia caicosensis*
(type specimen)
Photographic print
ca. 1997–2001
Photographer, William K. Sacco/Yale Peabody Museum
YPM IZAR 005918

85 *Tubipora* sp.
Peros Banhos Atoll, Chagos Archipelago, Indian Ocean
Photographer, Eric A. Lazo-Wasem/Yale Peabody Museum, 2005
YPM IZ 010514.CN
Collector, Willard D. Hartman, Yale Seychelles Expedition, October 23, 1957

86 Love nut (left)
Lodoicea maldivica
Seed
Praslin Island, Republic of Seychelles
YPM YU 101232
Collector, Willard D. Hartman, January 18–30, 1958

White oak acorn (right)
Quercus alba
Locality, unknown
Teaching collection

Chapter 3
From the Ends of the Earth

87 Detail of "Morrison Quarry 1"
Watercolor on paper
1878
Artist, Arthur Lakes (1844–1917), American
Photographer, Nathan Utrup/Yale Peabody Museum
YPM VPAR 002774

88 *Acropora carduus*
Fiji Islands
YPM IZ 001999.CNB
Collector, unknown, May 6–August 11, 1840, U.S. Exploring Expedition 1838-1842

Sketch of *Acropora carduus*
(YPM IZ 001999.CNA)
For James Dwight Dana, *Zoophytes.* Philadelphia: Lea and Blanchard, 1849, pl. 36, Figure 2
Graphite on paper
Undated
Artist, Alfred Thomas Agate (1812–1846)
YPM IZAR 005905

89 Sinnet line and hook
Plant material, shell
Samoa, Polynesia
19th century
Photographer, Rebekah DeAngelo/Yale Peabody Museum
YPM ANT 015001
Collector, unknown, U.S. Exploring Expedition 1838–1842

90 *Glossopteris browniana*
(type specimen)
New South Wales, Australia
Permian Period
YPM PB 008008
Collector, James D. Dana, 1839, U.S. Exploring Expedition 1838–1842

91 Fern
Hymenophyllum formosum (type specimen)
Tahiti, French Polynesia
Photographer, Division of Botany/Yale Peabody Museum
YPM YU 000642
Collector, unknown, 1839, U.S. Exploring Expedition 1838–1842

92 *Metacarcinus magister*
San Francisco Bay
Photographer, Eric A. Lazo-Wasem/Yale Peabody Museum, 2015
YPM IZ 000209.CR
Collector, Charles Pickering, ca. 1841, U.S. Exploring Expedition 1838–1842

93 Fossil horse teeth, vertebrae, fragments of limb bones
Protohippus parvulus (type specimen)
Kimball County, Nebraska
Miocene Epoch
YPM VP 011340
Collector, Othniel C. Marsh, 1868

94 Map, U.S. Army scouting party
Ink on translucent waterproof (?) paper
1870
YPM VPAR 002368
Donor, Lieut. Bernard Reilly, Jr.

95 Members of the Yale College Scientific Expedition of 1872
Standing, left to right: Guide Ed Lane, Othniel C. Marsh, and Lieut. James W. Pope
Seated, left to right: Thomas H. Russell ('72 S, M.D. '75), Benjamin Hoppin ('72), James MacNaughton, Charles D. Hill
Photographic print
1871
Photographer, unknown
YPM VPAR 000076

96 Rifle
Inscribed to Texas Charley (Charles Bigelow) by Buffalo Bill (William F. Cody)
Metal and wood
19th century
YPM ANT 013396
Donor, Mrs. C. Bigelow, 1923

97 Sketch of Edward Drinker Cope
From *Popular Science Monthly,* Vol. 19
May 1881
Artist, unknown

98 *Mosasaurus copeanus*
(type specimen)
Monmouth County, New Jersey
Cretaceous Period
YPM VP 000312
Collector, O.C. Herbert, 1868

99 Othniel C. Marsh and members of the Yale College Scientific Expedition of 1871
Photographic print (detail)
Near Salt Lake City, Utah
1871
Photographer, C. R. Savage
YPM YPMAR 000074

100 *Hadrosaur ilium* in matrix, shipping crate
YPM VPAR 000255
Collector, John B. Hatcher and crew, 1892, John Bell Hatcher Archives

101 Bingham Oceanographic Expeditions

Log book, R/V *Pawnee*
1925
YPM IZAR 005911

Specimens (left to right):

Red *sargassum*
Caribbean Sea
YPM IZ 076896
Collector, R/V *Atlantis,* February 23, 1934

Flying fish
Cypselurus vitropinna (type specimen)
Caribbean Sea
YPM ICH 000459
Collector, unknown, R/V *Pawnee,* February–April 1925

Red *sargassum*
Gulf of Mexico, Atlantic Ocean
YPM IZ 076894
Collector, unknown, R/V *Atlantis,* February 19, 1935

Lobster
Nephrops binghami
Caribbean Sea
YPM IZ 004382.CR
Collector, unknown, R/V *Pawnee,* April 4, 1925

Velvet whalefish
Barbourisia rufa (type specimen)
Gulf of Mexico
YPM ICH 001119
Collector, unknown, R/V *Atlantis,* March 28, 1937

Dragonfish
Echiostoma ctenobarba (type specimen)
Caribbean Sea
YPM ICH 002091
Collector, R/V *Pawnee,* March 1, 1927

Green *sargassum*
San Blas Islands, Panama, Caribbean Sea
YPM IZ 076895
Collector, G. C. Ewing, May 9, 1941

Water samples
Delaware Bay
ca. 1940s

102 Bear Seamount Expeditions specimens (left to right)

Lovely hatchetfish
Argyropelecus aculeatus
Bear Seamount, North Atlantic
YPM ICH 027674
Collectors, Gregory J. Watkins-Colwell, Tracey Sutton, and Jon A. Moore, R/V *Pisces,* October 15, 2014

Treadfin dragonfish
Echiostoma barbatum
Bear Seamount, North Atlantic
YPM ICH 025412
Collector, unknown, R/V *Pisces,* September 2, 2012

Barbeled dragonfish
Melanostomias bartonbeani
Bear Seamount, North Atlantic
YPM ICH 027508
Collectors, Gregory J. Watkins-Colwell, Tracey Sutton, and Jon A. Moore, R/V *Pisces,* October 21, 2014

Coral
Chrysogorgia tricaulis
Bear Seamount, North Atlantic
YPM IZ 038608
Collector, Les Watling, May 11, 2004

Coral
Paragorgia johnsoni
Bear Seamount, North Atlantic
YPM IZ 036781
Collector, Jon A. Moore (R/V *Delaware II*), June 9, 2004

Brittle star
Asteroschema sp.
Bear Seamount, North Atlantic
YPM IZ 076997
Collector, J.A. Moore (R/V *Delaware II*), June 9, 2004

103 Field register, Egypt
1962
Author, William K. Simpson
YPM ANTAR 035972
Donor, William K. Simpson, 1998

Kohl jar
Serpentine
Toshka, Nubia, Egypt
Middle Kingdom (2030–1650 BCE)
YPM ANT 265146
Simpson reg. no. 194
Collector, William K. Simpson, 1962

Cylindrical bead
Inscribed with representation of goddess Taweret on one side, hieroglyphs *Nbw-xpr-ra* (Antef V) on the other
Ivory
Toshka West, Nubia, Egypt
2nd Intermediate Period (1650–1550 BCE)
YPM ANT 265145
Simpson reg. no. 195
Collector, William K. Simpson, 1962

Wedjat eye bead
Faience, painted blue
Arminna West, Nubia, Egypt
Meroitic (270 BCE–350 CE)
YPM ANT 261936
Simpson reg. no. 197, field photo no. G30
Collector, William K. Simpson, 1962

William Kelly Simpson and excavation staff
Photographic print
Arminna West, Nubia, Egypt
1960
Photographer, unknown
YPM ANTAR 035582
Donor, William K. Simpson, 1998

Egyptian fruit bat (dry)
Rousettus aegyptiacus aegyptiacus
Aswan, Egypt
YPM MAM 014576
Collectors, Ibrahim Helmy and Christopher Maser, December 21, 1963

Nubian spitting cobra
Naja nubiae (type specimen)
Asyut, Egypt
YPM HERR 005210
Collector, Christopher Maser, December 11, 1963

Egyptian fruit bat (wet)
Rousettus aegyptiacus aegyptiacus
Aswan, Egypt
YPM MAM 014576
Collectors, Ibrahim Helmy and Christopher Maser, December 21, 1963

Common kestrel
Falco tinnunculus rupicolaeformis
Egypt
YPM ORN 096768
Collector, Thomas E. Lovejoy, November 6, 1962

104 Anchisaurus polyzelus
"Yaleosaurus" herbivorous dinosaur
Right forefoot and skull
Hartford County, Connecticut
Jurassic Period
YPM VP 001883
Collector, O.C. Wolcott, April 1891
Photographer, Jerry Domian/Yale Peabody Museum

105 A Collection of Plants of New Haven and its Environs Arranged According to the Natural Orders of Jussieu
One volume of specimens from a four-volume set
New Haven, Connecticut
Photographer, Division of Botany/Yale Peabody Museum
YPM BOTAR 000048
Collector, Horatio Nelson Fenn (1798–1871), 1822

106 Wolcott meteorite (type specimen)
Stony, ordinary chondrite
April 19, 2013 (recorded fall)
YPM MIN 100439
Donor, Darryl Pitt, March 6, 2014

Weston meteorite
(*see 9*)

"An Account of the Meteor"
(*see 9*)

Chapter 4
Conserving Nature

107 Army officers (*left to right*) Dr. Gandy, Lieutenant Nance, and Captain Scott at Fort Yellowstone, posing with buffalo heads likely confiscated from Ed Howell
Photographic print (detail)
ca. 1894
Photographer, unknown
YELL 7757
National Park Service Photo, Yellowstone National Park

108 Junction of the Platte Rivers
Oil on paper mounted on canvas
1866
Artist, Thomas Worthington Whittredge (1820–1910), American
Photographer, Yale University Art Gallery
YPM YPMAR 000742
Donor, William Farnam (B.A. 1866, M.A. 1869), 1929

109 Members of the Yale College Scientific Expedition of 1870
Standing left to right: John Reed Nicholson ('70), George Bird Grinnell ('70, Ph.D. '80), James Wolcott Wadsworth (M.A. '08), Othniel Charles Marsh ('60, M.A. '63), Charles Wyllys Betts ('67, M.A. '71), Harry Degen Ziegler ('71 S), Henry Bradford Sargent ('71 S, M.A. '07)
Seated, left to right: John Wool Griswold ('71 S), Alexander Hamilton Ewing ('69), Eli Whitney ('69, M.A. '72), Charles McCormick Reeve ('70, M.A. '73), James Matson Russell ('70)
Photographic print
Chicago, Illinois
1870
Photographer, unknown
YPM VPAR 000072

110 Greater roadrunner
Geococcyx californianus
Los Angeles, California
YPM ORN 144270
Collector, George B. Grinnell, March 1876

111 "On the Osteology of *Geococcyx californianus*"
Handwritten dissertation draft, select pages
Author, George B. Grinnell (1849–1938)
Sterling Memorial Library, Yale University, MS 1388, Accession 2005-M-033

112 George Bird Grinnell on Grinnell Glacier
Photographic print
1925
Photographer, T. J. Hileman
Courtesy of Glacier National Park Archives

113 Grinnell Glacier over time
Grinnell Glacier, ca. 1887 (left)
Photographic print
Photographer, Grinnell's assistant
Glacier National Park Archives

Grinnell Glacier, 2013 (right)
Photographer, Dan Fagre, USGS
U.S. Geological Survey Repeat Photography Project (nrmsc.usgs.gov/repeatphoto/)

114 The Audubon Magazine
Volume 1, no. 1, February 1887 (reprint)
YPM YPMAR 000745

115 George Bird Grinnell Roosevelt Memorial Association Distinguished Service Medal
Gold
1925
YPM ANT 206133
Donor, Eliz. Grinnell Trust, 1961

116 Earthrise
70 mm magazine B
December 24, 1968
Photographer, William A. Anders
Image Science and Analysis Laboratory, NASA – Johnson Space Center, image AS08-14-2383

117 Annual meeting program, Xerces Society
Mimeographed sheet
April 1974
YPM ENTAR 001709
Donor, Charles L. Remington, 1974

Announcement, Xerces Society
Postcard
1972
YPM ENTAR 001709
Donor, Charles L. Remington, 1974

Xerces Blue butterflies
Glaucopsyche lygdamus xerces
San Francisco County, California
YPM ENT 447194
Collector, C.A. Hill, March 14, 1928

118 Charles L. Remington with hissing cockroaches
Photographic print
Osborn Memorial Laboratory, New Haven
January 1966
Photographer, unknown
YPM ENTAR 002717

119 Gynandromorph butterflies
Daxocopa laurentia
YPM ENT 844629 (male, top left)
Pelotas, Brazil
Collector, C. Biezanko, December 14, 1954

YPM ENT 844637 (female, top right)
Pelotas, Brazil
Collector, C. Biezanko, December 14, 1954

YPM ENT 745182 (gynandromorph, bottom)
Nova Teutonia, Brazil
Collector, Fritz Plaumann, January 15, 1958

120 Gynandromorph butterflies
Lycaena gorgon
YPM ENT 844634 (female, left)
Keddie, California
Collector, Frank Morton Jones, June 15, 1918

YPM ENT 844636 (gynandromorph, middle)
Castle Rock Park, California
Collector, Paul Opler, May 10, 1953

YPM ENT 844639 (male, right)
Santa Ynez Mountains, California
Collector, Carl Kirkwood, June 6, 1938

121 17-year cicada, Broods II and XI
Magicicada septendecim
New Haven County, Connecticut
1843, 1860, 1877
YPM ENTAR 002778

Collectors, Cornelia L. Hillhouse, Jerauld A. Manter, Oscar Harger, and Benjamin Silliman

122 Verreaux's sifaka, or white sifaka
Propithecus verreauxi
Kirindy Mitea National Park, Madagascar
Undated
© Dennis van de Water/Shutterstock.com

123 Beza Mahafaly team members
Digital photograph
2011
Photographer, Roshna Wunderlich
Used with permission

124 Beza Mahafaly team members
Digital photograph
2011
Photographer, Roshna Wunderlich
Used with permission

Chapter 5
Evolving Science

125 Deinonychus antirrhopus
Mounted skeleton, composite cast
Carbon County, Montana
Cretaceous Period

126 "Restoration of (Ichthyornis) Dispar"
Ink on paper
1880
Artist, Thomas Nast (1840–1902)
YPM VPAR 002242

127 Ichthyornis dispar
(type specimen)
Cast of YPM VP 001450
Scott County, Kansas
Cretaceous Period
YPM VP 005709
Collector, Benjamin F. Mudge, 1872

128 Ichthyornis dispar
(type specimen)
Lower mandible with teeth
Scott County, Kansas
Cretaceous Period
YPM VP 001450
Collector, Benjamin F. Mudge, 1872

129 Lecture room in the first Peabody Museum building
Photographic print
ca. 1878
Photographer, unknown
YPM VPAR 000345

130 Letter from Charles Darwin to O. C. Marsh
August 31, 1880
YPM YPMAR 000895
Othniel Charles Marsh Papers, Yale Peabody Museum Archives

Odontornithes: A monograph on the extinct toothed birds of North America
Memoirs of the Peabody Museum of Yale College, vol. 1, New Haven, 1880
Author, Othniel Charles Marsh (1831–1899)
YPM VPAR 002771

131 Mounted skeleton
Hesperornis crassipes
Gove County, Kansas
Cretaceous Period
YPM VP 001474
Collectors, George P. Cooper, Benjamin F. Mudge and party, Yale 1876 Kansas Expedition, April 22, 1876

132 Mounted foot, cast
Deinonychus antirrhopus
Carbon County, Montana
Cretaceous Period
YPM VP 058123 (mounted cast of left pes), YPM VP 005205
Collector, unknown, Yale Cloverly Expeditions, 1964–1965

133 Restoration of a hypothetical proavian
From *Vor Nuvaerende Viden om Fuglenes Afstamming*, Danish edition, 1916
English edition, *The Origin of Birds*. London, H. F. & G. Witherby, 1926
Author, Gerhard Heilmann (1859–1946)

134 John H. Ostrom
35 mm color slide
1962
Photographer, unknown
Courtesy of the Ostrom family

135 Archaeopteryx lithographica
Photographic print
ca. 1972
Photographer, John H. Ostrom
YPM VPAR 002493

136 Half-moon wrist bone (carpal)
Deinonychus antirrhopus
Carbon County, Montana
Cretaceous Period
YPM VP 005208
Collector, Peter Parks, Cloverly Expedition, 1965

137 Reconstruction of *Anchiornis huxleyi*
Watercolor on paper
Painting by Michael DiGiorgio; mdigiorgio.com

138 Pompadour cotinga, male (top)
Xipholena punicea
French Guiana
YPM ORN 029949
Collector, S. M. Klages, May 17, 1917

Plum-throated cotinga, male (bottom)
Cotinga maynana
Peru
YPM ORN 081770
Collector, Celestino Kalinowski, July 10, 1957

139 Experimental restoration of predicted ancestral expression
Composite digital photograph
From "A molecular mechanism for the origin of a key evolutionary innovation, the bird beak and palate, revealed by an integrative approach to major transitions in vertebrate history." *Evolution* 69(7), Figure 4
Authors, Bhart-Anjan S. Bhullar, Zachary S. Morris, Elizabeth M. Sefton, Atalay Tok, Masayoshi Tokita, Bumjin Namkoong, Jasmin Camacho, David A. Burnham, and Arhat Abzhanov
July 2015
Used with permission

140 Fossilized feathered skeleton of sandpiper (part and counterpart)
cf. Scolopacidae
Madison County, Montana
Oligocene Epoch
YPM VPPU 017053
Collector, Herman F. Becker, August 25, 1959

Chapter 6
One in a Million

141 Georg Hartman astrolabe
Brass
Nüremberg, Germany
1537
YPM HSI 040001
Donor, Hewitt Cochran (B.A. '38), 1972

142 Inflated Lepidoptera caterpillars
Various species and localities
1912 to 1933
YPM ENTAR 002784
Donor, United States Forest Service, 1988

143 Structural coloration in insects (drawer)
Various localities
1897 to 1980s
YPM ENTAR 002782
Collectors, Eugene Dluhy, Celestin Crozet, Charles Bridgham, et al.

144 Point-tailed palmcreeper nest
Berlepschia rikeri
Republic of Suriname
YPM ORN 137643
Collector, Kristof Zyskowski, June 2007

145 Quagga mare
Equus quagga quagga (extinct)
Photographic print
1870
Photographer, Frederick York (1823–1903)
Zoological Society of London

146 Skull, female
Equus quagga quagga (extinct)
South Africa
YPM MAM 007481
Collector, Edward Gerrard, Jr., 1872

147 Uintacrinus socialis
(type specimen)
Logan County, Kansas
Cretaceous Period
YPM IP 024758
Collector, Handel T. Late Martin, 1898

148 Metasequoia occidentalis
Madison County, Montana
Oligocene Epoch
YPM PB 035960
Collector, Herman F. Becker

149 Ginkgo adiantoides
Dawson County, Montana
Cretaceous Period
YPM PB 007004
Collector, Erling Dorf

150 Reconstruction of eurypterid
Digital illustration
2008
Artist, Simon Powell
Used with permission

151 Acutiramus macrophthalmus
Herkimer County, New York
Silurian Period
Photographer, Jessica Utrup/Yale Peabody Museum
YPM IP 208194
Collector, S. J. Ciurca, March 26, 1967

152 Crocodylomorpha
Garfield County, Utah
Triassic Period
YPM VP 057103
Collector, Yale Vertebrate Paleontology Field Party, 2003, 2005

153 Blackfin icefish
Chaenocephalus aceratus
Southern Ocean, Antarctica
YPM ICH 024284
Collector, Kristen L. Kuhn, May 14, 2010

154 Giant isopod
Bathynomus giganteus
Gulf of Mexico
Photographer, Eric A. Lazo-Wasem/Yale Peabody Museum
YPM IZ 078000
Acquired by Gregory J. Watkins-Colwell, R/V *Pisces*, August 2014

155 Giant clam
Tridacna gigas
Locality unknown
YPM IZ 024044
Photographer, unknown

156 Double cylinder electrical machine
Wood, glass, sealing wax, brass, silk, tin foil
ca. 1770
YPM HSI 010050

157 Vacuum chamber
27-inch cyclotron
Made by Edward O. Lawrence
1939
YPM HSI 290013

158 Charles R. Darwin (1809–1882)
Photographic print
Undated
Library of Congress Prints and Photographs Division, Washington, D.C.
Reproduction no. LC-USZ61-104

159 Lively liverwort
Lepidozia chordulifera
Chonos Archipelago, Chile
YPM YU 091382
Collector, Charles R. Darwin, December 1834

160 Sea fan
Lophogorgia sanguinolenta
Locality not specified
Photographer, Eric A. Lazo-Wasem/Yale Peabody Museum
YPM IZ 000952.CN
Received from Charles Darwin, date unknown

161 "HMS Beagle in the Straits of Magellan"
Frontispiece from *Journal of researches into the natural history and geology of the various countries visited by H.M.S. Beagle etc.*
First Murray illustrated edition, 1890
Author, Charles Darwin
Artist, Robert Taylor Pritchett (1828–1907)

162 Thomas H. Huxley
Line engraving
1874
Artist, C. H. Jeens
Wellcome Library, London

163 "*Eohippus & Eohomo*"
Ink on paper
August 1876
Artist, Thomas H. Huxley (1825–1895)
YPM VPAR 002503

Fossil Horses in America
[Salem, Salem Press], 1874
Author, Othniel C. Marsh
Owned by George R. Wieland
YPM VPAR 000896

164 Debating stool
Painted wood, inlaid shell
Lake Chambri Region?, Middle Sepik, Papua New Guinea
Mid-20th century
YPM ANT 250091
Collectors, Dr. Ruth Lidz and Dr. Theodore Lidz, 1986

165 Michael D. Coe, Monument 34, San Lorenzo, Tenochtitlan, Veracruz, Mexico
35mm color slide
1960s
Photographer, Raymond Krotser
Courtesy of Michael D. Coe

166 Colossal Head 6, Olmec
Plastic, cast
1200–900 BCE
YPM ANT 264662
Collector, Michael D. Coe, 1966–1968 San Lorenzo

167 First Herbarium of Amos Eaton
One volume of specimens from a four-volume set
New York and New England
Photographer, Division of Botany/Yale Peabody Museum
YPM BOTAR 000317
Collector, Amos Eaton (1776–1842), ca. 1816

168 Ferns of North America
Vol. 1, p. 293, plate 39
Salem, Mass.: S. E. Cassino, Naturalists' Agency, 1879–1880
Author, Daniel Cady Eaton (1834–1895), drawings by J. H. Emerton and C. E. Faxon

169 Rice-beer jar
Porcelain
Province of Ifugao, Philippines
Ming Dynasty, China (1300–1600 CE)
Photographer, William K. Sacco/Yale Peabody Museum
YPM ANT 262793
Collector, Harold C. Conklin, 1984

170 Exhibit display, Gargantua
Postcard
Publisher, The Meriden Gravure Company, Meriden, Conn., for the Peabody Museum
1951
Gorilla gorilla
Carbon County, Montana
YPM MAM 009752

171 "Ringling Bros. and Barnum & Bailey: Gargantua the Great"
Circus poster
Ink on paper
Undated
The Strobridge Lithographing Company
Attributed to Howard Sharp (1878–1952)
Collection of the John and Mable Ringling Museum of Art, Tibbals Digital Collection

172 Ivory-billed woodpecker
Campephilus principalis principalis (probably extinct)
Adult male (left)
Florida
YPM ORN 004632
Collector, P. L. Jouy, before 1894

Immature female (right)
Osceola County, Florida
YPM ORN 004633
Collector, Thomas A. James, January 15, 1890

173 Krasnojarsk meteorite
Stony-iron, pallasite
Krasnoyarsk Territory, Russia
Found 1749
YPM MIN 100388
Gibbs Cabinet, purchased 1825

174 Widmanstätten pattern
Digital photograph
April 4, 2015
Photographer, Ivtorov/CC-BY-SA 4.0 International

175 Tyrannosaurus rex tooth
Jefferson County, Colorado
Cretaceous Period
Photographer, Jerry Domian/Yale Peabody Museum
YPM VP 004192
Collector, Peter T. Dotson, 1874

Chapter 7
A Changing Earth

176 Sabalites sp. with *Priscacara* sp. (large) and *Knightia* sp.
Lincoln County, Wyoming
Eocene Epoch
YPM PB 168197
Purchased, 2013

177 Alligator cf. *prenasalis*
Custer County, South Dakota
Oligocene Epoch
Photographer, Jerry Domian/Yale Peabody Museum
YPM VPPU 013799
Collector, J. Clark, August 15, 1933

178 Metacarbonate
Marlas, Greece
YPM MIN 100577
Collector, Jay J. Ague, June 30, 2001

179 Three-pointed zemi
Stone
Dominican Republic
Boca Chica period (ca. 1200–1500 CE)
YPM ANT 024697
Collector, Armour-Caribbean Expedition, 1934

180 Pendant
Shell
Haiti
Couri period (2000 BCE–500 CE)
YPM ANT 011444
Collector, Irving Rouse, 1935

181 Saladoid series white-on-red bowl
Ceramic
St. Croix, Virgin Islands
Ceramic age (250 BC–500 CE)
YPM ANT 237762
Collector, Fred Olsen Collection, 1982

182 Green tree frog
Rana clamitans melanota
New Haven County, Connecticut
YPM HERA 005800
Collector, Walter Gradzik, John Gradzik, Walter Gradzik, Jr., July 2000

183 "The City of New Haven, Conn. 1879"
Map, 66 × 96 cm (26 × 37.8 in.)
Published, Boston
ca. 1879
Artists, O. H. Bailey (1843–1947) and J. C. Hazen
Library of Congress, Geography and Map Division, G3784.N4A3 1879.B3

184 Pyrola elliptica in flower
Digital photograph
© Arthur Haines/New England Wild Flower Society
Used with permission

185 Elliptic-leaved shinleaf
Pyrola elliptica
Sachem's Wood, New Haven County, Connecticut
Photographer, Division of Botany/Yale Peabody Museum
YPM YU 058303
Collector, James D. Dana, July 1851

186 Asian shore crab
Hemigrapsus sanguineus
Fairfield County, Connecticut
Photographer, Eric A. Lazo-Wasem/Yale Peabody Museum
YPM IZ 078539
Collector, Strength in Numbers Workshop, July 16, 2013

187 Black oak
Quercus velutina
New Haven County, Connecticut
Photographer, Division of Botany/Yale Peabody Museum
YPM YU 056031
Collector, Martha Hill, August 7, 1985

page 114 Tapa cloth
Inner bark of mulberry tree
Mid 20th century
Samoa, Polynesia
Brandt collection
YPM ANT 248711

Further Reading

Benson, Richard. *A Yale Album: The Third Century.* New Haven: Yale University Press, 2000.

Conniff, Richard. *House of Lost Worlds: Dinosaurs, Dynasties, and the Story of Life on Earth.* New Haven: Yale University Press, 2016.

Creating the Peabody's Torosaurus: *Dinosaur Science, Dinosaur Art.* Dir. Ann Johnson Prum. Yale Peabody Museum, 2006. DVD.

Lavin, Lucianne. *Connecticut's Indigenous Peoples: What Archaeology, History, and Oral Traditions Teach Us About Their Communities and Cultures.* Ed. Rosemary Volpe. New Haven: Yale Peabody Museum of Natural History and Yale University Press, 2013.

Logan, Alison M.B. "A Museum of Ideas: Evolution Education at the Peabody Museum during the 1920s." Yale Peabody Museum of Natural History. Online article. peabody.yale.edu

McCarren, Mark J. *The Scientific Contributions of Othniel Charles Marsh.* New Haven: Peabody Museum of Natural History, Yale University, 1993.

Migdalski, Ed. *Lure of the Wild: The Global Adventures of a Museum Naturalist.* New Haven: Peabody Museum of Natural History, Yale University, 2006.

Ostrom, John H., and John S. McIntosh. *Marsh's Dinosaurs: The Collections from Como Bluff.* 2nd ed. New Haven: Yale University Press, 1999.

Parker, Franklin. *George Peabody: A Biography.* 1971. Nashville: Vanderbilt University Press, 1995.

Prince, Cathryn J. *A Professor, a President, and a Meteor: The Birth of American Science.* Amherst, NY: Prometheus Books, 2011.

Scarisbrick, Diana and Benjamin Zucker. *Elihu Yale: Merchant, Collector and Patron.* New York: Thames & Hudson, 2014.

Thomson, Keith. *The Legacy of the Mastodon: The Golden Age of Fossils in America.* New Haven: Yale University Press, 2008.

Volpe, Rosemary, ed. *The Age of Reptiles: The Art and Science of Rudolph Zallinger's Great Dinosaur Mural at Yale.* 2nd ed. New Haven: Peabody Museum of Natural History, Yale University, 2007.

Wallace, David Rains. *Beasts of Eden: Walking Whales, Dawn Horses, and Other Enigmas of Mammal Evolution.* Berkeley: University of California Press, 2005.

Index

K

L

M

N

O

P

Q

R

S

T

U

V

W

X

Y

Z